£7-50

**This book is to be returned on or before
the last date stamped below.**

Ideas, concepts and problems which challenge
the mind and baffle the experts

Think of a Number

Malcolm E Lines

Adam Hilger, Bristol and New York

British Library Cataloguing in Publication Data

Lines, M.E. (Malcolm E)
    Think of a number.
    1. Mathematics. Problem solving
    I. Title
    510

    ISBN 0-85274-183-9

Library of Congress Cataloging-in-Publication Data

Lines, Malcolm E.
    Think of a number: ideas, concepts, and problems which challenge the mind and baffle the experts/Malcolm E. Lines.
        p.   cm.
    Published also under title: Numbers at work and at play.
    Includes bibliographical references.
    ISBN 0-85274-183-9
    1. Mathematics—Popular works.   I. Title.
    QA93.L554   1990   89-24681
    510—dc20   CIP

Published under the Adam Hilger imprint by IOP Publishing Ltd
Techno House, Redcliffe Way, Bristol BS1 6NX, England
335 East 45th Street, New York, NY 10017-3483, USA

Filmset by Bath Typesetting Ltd, Bath
Printed in Great Britain by J W Arrowsmith Ltd, Bristol

Contents

Preface

One morning, back in the spring of 1961, I found myself sitting at the end of a truly impressive oak table in the Summer Common Room of Magdalen College, Oxford, defending some of my research work before the Fellows of the College in the oral part of a Fellowship examination. Since I was a physicist, with some mathematical leanings, most of the questioning came from the scientists and mathematicians present. These questions centred for a while on some rather arcane mumbo-jumbo about mathematical objects known as 'Green's functions' which, at the time, were rather in vogue in my immediate field of theoretical research, but represented hardly more than voodoo mathematics to non-specialists—even among scientists. In fact, to be quite honest, most scientists of the day had probably never heard of them. Nevertheless to me, at the time, they were important and I felt that I was fielding the questions quite well.

It was at this moment (by which time I was beginning to grow a little in confidence) that the questioning was opened up to the audience at large, most of whom, though eminent in their own fields, were not scientifically oriented and had almost certainly been struggling to stay awake during the mathematical ramblings of the preceding forty minutes. One of them, a long-serving Fellow of the College, resplendent in gown, and fixing me with a piercing glare, rose slowly to his feet. Evidently annoyed by the fact that all the prior discourse had been utterly unintelligible to him (and presumably to most of the others present) he posed a question which haunts me to this day. 'These Green's functions that I hear you talking so much about,' he said, 'how would you explain one of those to a medieval historian?' The fact that I recall the question word for word to this day, without having any recollection whatsoever of my answer, probably speaks eloquently for the quality of the response.

Several years later, I found my way to Bell Research Laboratories, New Jersey, USA, where I plied my trade as a solid state physicist. This was the period of the computer revolution, with the company purchasing bigger and faster 'number-crunchers' every few years, making it ever more convenient to think less and compute more. Occasionally, however, when I had a mathematical problem for which I felt it likely that the equations possessed *exact* (or what mathematicians call *analytic*) solutions, I would resist looking for numerical answers on the computer and go to the office of one of my older colleagues, a kind and gentle man who had had the good fortune to mature scientifically in the years when thinking was less avoidable. 'I feel sure that these equations have an analytic solution,' I would say, 'but I can't seem to find it. Am I being stupid?' 'Perhaps, just a little', he would often respond with an understanding half-smile, before leading me gently in the direction of the proper solutions.

One afternoon, in the summer of 1980, while on just such a mission, I found him to be uncharacteristically effervescent. He jumped to his feet and, before I could ask my customary question, thrust some papers into my hand. 'Read this,' he said, 'it is some of the most fascinating work that I have ever seen; wonderfully profound but so elegantly simple.' I resisted drawing the unintended inference—that its simplicity was of a degree that even I (or dare I say a medieval historian) might appreciate its essentials. It was, in fact, the early work on the theory of the onset of chaos, about which you will learn more later, should you decide to read on in this book.

During this same period of time, my wife and I would socialize about once a month with a younger couple who lived a few doors away. He was a builder and she a housewife and part-time designer–dress distributor. The evenings were always relaxing and pleasant and the conversation not particularly academic. In fact, we would often while away the hours half-playing Mah-Jongg (a Chinese game with tiled pieces which, I am told, was all the rage in the 1920s) while simultaneously recounting any worthwhile anecdotes pertaining to our experiences since we had last met. And then, on one occasion, without any interruption in the flow of the conversation, my hostess surprised me by saying 'I hear that you are thinking of writing a book about numbers. Are you going to say anything about the Fibonacci's?'

My purpose in recalling these 'verbal snapshots' from the past is not, of course, to try to suggest that historians of any kind secretly thirst for knowledge about Green's functions, nor that the new and fascinating field of chaotic motion can be appreciated in all its details by the completely uninitiated. It is to make three separate points. Firstly, that someone who claims to understand, and be excited by, any aspect of science (and yes, even mathematics) ought to be able to pass on the essence of that knowledge and enthusiasm to any reasonably intelligent layperson who is interested. Secondly, that many of the most exciting advances of this kind do lend themselves admirably to just such exposition. And finally, and perhaps most importantly, that there may be a much wider potential interest 'out there' than anyone suspects–if only authors would make a serious effort to bridge the verbal chasm between the specialized jargon of the learned journals and the normal vocabulary of the population at large. This book is a modest effort to encourage such a trend.

Malcolm E Lines
August 1989

1

Introduction

Throughout the ages, ever since man first acquired an interest in counting and measuring, the concept of 'number' has gradually developed to fascinate and sometimes torment him. From the simplest ideas concerning the familiar 1,2,3, through negative numbers, to fractions, decimals and worse, the basic understanding of what one ought to mean by 'number' in its most general sense steadily increased. And growing with it in an equally relentless fashion was a set of fascinating questions and speculations concerning the many weird and wonderful properties of these numerical notions. Some of the related problems were quickly 'cleared away' to the satisfaction of the experts of the day. Others yielded after much longer periods of effort—sometimes decades, and occasionally even centuries. A few live on in infamy, and continue to baffle the world's greatest mathematicians (with or without the assistance of their powerful latterday allies, the electronic digital computers) and to test their ingenuity and sanity.

Evidently, by its title, this book is about numbers in some sense. But this time not so much about the properties of numbers themselves (which have already been probed in the companion book *A Number for your Thoughts*) as of the interplay of numbers with 'nature' in a very general sense. Some of the examples seem, outwardly at least, to be of a lighter vein; involving hailstorms, taxi-cabs, patio decor, pine cones, bicycle assembly and colouring books (numbers at 'play', if you will). Others are concerned with seemingly weightier topics such as secret codes and national security, symmetry and atomic physics, meteorology, the bending of space in the fourth dimension (or even the $3\frac{1}{2}$th) and information network systems. Here, the interaction of numbers and the problems of the real world seems to have more serious consequences—one perhaps more akin to 'work'. In fact, I was tempted to entitle the book 'Numbers at Work and at Play' at one time—but it sounded too much like an elementary arithmetic book for pre-school children,

1

and that it most certainly is not. Each story told in this book, whether it
depicts numbers at work or at play according to my definition, gives insight
(in what I hope is an entertaining fashion) into problems involving deep-
down mathematical notions, often with very important consequences. Many
pursue ideas which have evolved over centuries of study, others are of
extremely recent origin. Some are now fully understood, others are now only
at the very beginning of their development. Some have required the 'number-
crunching' impact of today's most powerful computers to reveal their secrets,
others have yielded to an inspired moment of pure thought.

In spite of taking us to the 'cutting-edge' of today's research in many
instances, our stories require very little knowledge of mathematics to
understand them. In fact, anyone who remembers even one half of his first
year's high-school algebra can happily skip the rest of this introduction and
move right along to 'The Fibonacci Saga' of Chapter 2. For the rest of you it
is perhaps a good idea to go over the meaning of a few words that might
appear without explanation in the text (and which may not have been a
regular part of your recent mathematical conversation!).

Firstly, the counting numbers (that is, the whole numbers 1,2,3,4, and so
on) are referred to as *integers*. Those integers which can be divided exactly (or,
in other words, without remainder) only by themselves and by 1, are called
prime numbers. Examples might be 3, or 11, or 29. All the other integers are
then said to be *composite*. It follows that all composite numbers can be formed
by multiplying together smaller integers; for example, $32 = 4 \times 8$. These
smaller integers are called *factors* of the larger one, and one set of factors is
rather special, namely the *prime factors*. Since neither 4 nor 8 in the example
above is a prime number, each can be 'factorized' further (and possibly further
still) until eventually only prime numbers are left. For the particular case of the
composite number 32 this happens when we reach

$$32 = 2 \times 2 \times 2 \times 2 \times 2.$$

These five twos are the prime factors of 32, and prime factors are special
because every composite number has one, and only one, such set. The prime
numbers therefore represent, in a way, the atoms (or smallest parts) from
which all other numbers are uniquely formed by multiplication. We say that
all integers are the *product* of their prime factors. Product therefore means
multiplication. Thirty-two is the product of 8 and 4 as well as of five twos.
The equivalent term for addition is *sum*, as in 32 is the sum of 28 and 4.

One other thing which we notice about the equation above is that it does
not look very elegant with all those twos on the right-hand side. We should
really be in trouble if our composite number called for say 50, rather than five,
twos. To deal with this, mathematics has invented a shorthand in which the
above equation is restated as

$$32 = 2^5.$$

The superscript 5 in this form is called a *power* or *exponent*, and tells you how many twos to multiply together. Another example might be

$$8 = 2^3.$$

By forming the product of 32 and 8 (i.e., $32 \times 8 = 256$) and factorizing the result as $256 = 2^8$, we can now write

$$2^5 \times 2^3 = 2^8$$

and notice something important. It is that when we multiply together numbers which are written in this way, we add the exponents ($5 + 3 = 8$).

Among other things this shorthand enables us to write down statements involving extremely large numbers and know that they are true, even though our pocket calculators (or even large computers) have no idea what these numbers look like when written out in full (or in *decimal notation* as the mathematician would say). We know, for example, that $317^{12} \times 317^{12} = 317^{24}$ in spite of the fact that the number on the right-hand side contains no less than 61 digits in decimal notation. But moreover, since 317^{12} is just a number like any other it can, when multiplied by itself, also be expressed in the new shorthand form as $(317^{12})^2$. It therefore follows that

$$(317^{12})^2 = 317^{24}$$

from which we learn the rule that a power raised to another power gives a new exponent which is just the product of the two originals (i.e., $12 \times 2 = 24$).

Since 317^{12} means twelve 317s multiplied together, it is quite obvious what 317^n implies so long as n is a positive integer. It is true that for 317^1, the literal extension to 'one 317 multiplied together' sounds a bit odd, but it must be equal to 317 because, only in this way would

$$317^1 \times 317^1 = 317^2$$

make sense using our power-addition rule ($1 + 1 = 2$). And what about 317 to the power zero? In words it should be 'no 317s multiplied together' and, if anything ought to be equal to zero, surely this should. But it is not! This we know by again using our one trusty power-addition rule in the form

$$317^0 \times 317^1 = 317^1$$

which has to be true since the exponents, or powers, add up correctly in the form $0 + 1 = 1$. And since 317^1 is just 317, as set out above, this equation becomes

$$317^0 \times 317 = 317$$

and can only be true if the zeroth power of 317 is equal to 1. Now, of course, the number 317 plays no significant role in all of this and I could have used any other number in its place with the same result. It follows that any number raised to the zeroth power is equal to 1. Did I say *any* number? Well, nearly any number; there is a little bit of trouble with that most unlikely looking number 0^0, but I will return to that in a moment.

First let us ponder the question of what some number raised to a fractional power means. Once again, our power-addition rule comes to the rescue. Using it, it must be true that

$$2^{1/2} \times 2^{1/2} = 2^1 = 2$$

and

$$2^{1/3} \times 2^{1/3} \times 2^{1/3} = 2^1 = 2$$

because, in each case, the powers add up properly. Thus, 2 to the one-half power is just that number which, when multiplied by itself, makes 2 (i.e., it is the *square root* of 2 or, on my calculator, 1.414 213 562...). In the same way 2 to the one-third power is the quantity which when multiplied by itself, and then by itself again, makes 2; this is the *cube root* of 2 or 1.259 921 050.... Just as simply $2^{1/10}$ or equivalently $2^{0.1}$ is the tenth root of 2 and so on. Fractions which do not have a 1 on the top are no more difficult in principle. For example,

$$3^{2/3} \times 3^{2/3} \times 3^{2/3} = 3^2 = 9$$

tells us immediately that $3^{2/3}$ is the cube root of 9.

Negative powers can also be understood via such equations as

$$2^1 \times 2^{-1} = 2^0 = 1$$

from which we see that 2^{-1} must be equal to 1/2. In fact, $2^{-n} = 1/(2^n)$ for any n. Negative numbers do occasionally get us into trouble on our calculators. For example, if you have a key labelled y^x on your pocket calculator and 'punch in' values $y = -1$, $x = 0.5$ (asking for the square root of -1), it will say something like DATA ERROR. This message of despair is telling us that the answer cannot be given in terms of the 'everyday numbers' like 1.8 or -2.7 which calculators can understand. It requires so-called '*complex numbers*' about which, mercifully, neither you nor your calculator need worry while reading this book.

But this still leaves us with 0^0. Entering $y = 0$ and $x = 0$ via the y^x button into my trusty Hewlett–Packard still produces that cry of frustration 'DATA ERROR'. With no negative numbers involved the problem cannot possibly involve complex numbers. So why can the calculator not give me an answer? Although of no direct relevance to the topics discussed in this book, it is a useful exercise in exponents to find the answer. Suppose first that I let $y = x = n$, and make n get smaller and smaller; 0.001, 0.000 01, and so on. Via my y^x button I now get answers (0.9931..., 0.9998..., and so on) which get closer and closer to 1.

So why is 0^0 not equal to 1? The answer is best understood by thinking of x and y as the axes on a piece of graph paper with $x = y = 0$ at its centre (or *origin*). We have so far only considered approaching the origin along the line $x = y$. If we approach from other directions (say along the x-axis, with $y = 0$ and x non-zero but getting smaller and smaller) we get other answers (like zero). The number 0^0 therefore has a *limiting* value which depends on how you approach it. The limiting value can be anything between 0 and 1 depending on the manner in which x and y are related as they both get smaller and smaller. Only very special ways of approaching the limit produce answers other than 1. In a sense, therefore, 0^0 is 1 unless you are unlucky! This is a true but painfully unmathematical statement. More precisely, to obtain a limiting value for 0^0 different from 1, it is necessary to approach the origin of the piece of $x-y$ graph paper closer to the x-axis than any power of x, or (for the experts) logarithmically close.

2

The Fibonacci Family
and Friends

The so-called Hindu–Arabic system of numbers 1,2,3,4,... was first spread into Europe primarily by the publication of certain books which both introduced them and demonstrated their many advantages over the older systems. By far the most influential of these was a book called *Liber Abaci* (which translates to 'a book about the abacus'), written by a remarkable Italian mathematician Leonardo Fibonacci. This book, written by the then 27 year old Fibonacci (whose surname literally means 'son of Bonacci') in the year 1202, has survived to this day in its second edition, which dates from 1228.

Now *Liber Abaci* is a book of considerable size, and records within its covers a large fraction of the known mathematics of those times. In particular, the use of algebra is illustrated by many examples of varying degrees of difficulty and importance. Strangely, one and one alone has achieved a fame far beyond the others. It is found on pages 123–4 of the surviving second edition of 1228 and concerns the unlikely problem of breeding rabbits. In essence it poses the following question: how many pairs of rabbits can be produced from a single pair in one year if every month each pair produces one new pair, and new pairs begin to bear young two months after their own birth? There is here, presumably, a subtle assumption that every pair referred to is composed of a male and a female (a condition which severely strains the laws of probability) but setting that aside as biology rather than mathematics, the remaining computation is not a difficult one. With a little thought one can easily derive the build-up of population to obtain the following sequence of numbers which counts the numbers of rabbit pairs munching on their food in each of the calendar months between January (when the first infant pair was introduced) and December:

$$1, 1, 2, 3, 5, 8, 13, 21, 34, 55, 89, 144$$

Looking at this sequence we soon observe that it is made up in an extremely simple fashion which in words may be stated as follows: each number (except, of course, the first two) is composed of the sum, which means addition, of the two preceding ones. Thus, for example, at the end of the above sequence, the December number 144 is obtained by simply adding together the October and November numbers 55 and 89.

This is all very well I suppose, but why on earth should it create any excitement even among mathematicians (whose threshold for jubilation often confounds the layman)? If this chapter serves its intended purpose, at least a few of the weird and wonderful properties which are spawned by these 'Fibonacci numbers' should make their fascination more understandable. But first let us generalize the sequence. Ignoring the mortality of rabbits, or even any decline with age in their ability to reproduce with this clockwork regularity, it is quite clear that the above list of numbers can be continued indefinitely to ever larger quantities. Indeed, we can even forget all about rabbits and just define the whole list as the infinite series of numbers for which the nth member, which we write as F_n (with the F in honour of Fibonacci), is simply defined as the sum of the two preceding numbers F_{n-1} and F_{n-2}. In this form the series of Fibonacci numbers is written as

$$F_1, F_2, F_3, F_4, F_5, ..., F_n, ...$$

where the dots imply a continuation *ad infinitum*. Using this notation we can now write down the very simple equation which, once $F_1 = 1$ and $F_2 = 1$ are given, enables us to determine all the subsequent numbers in turn: it is

$$F_n = F_{n-1} + F_{n-2}.$$

If $n = 3$ we obtain $F_3 = F_2 + F_1 = 1 + 1 = 2$. In a similar way we find $F_4 = 3, F_5 = 5$ and so on, the defining equation being valid for any value of n greater than or equal to 3.

As the Fibonacci numbers are continued beyond the value $F_{12} = 144$, which was the largest set out in the original rabbit problem, they begin to grow quite rapidly. For example, the 25th member of the series is already 75 025 while the 100th member, F_{100}, is a whopping

354 224 848 179 261 915 075

with 21 digits. Moreover, as they grow they are in a sense settling down into an even simpler pattern than their defining equation would at first sight suggest. This pattern is most easily recognized if we write down the ratio formed when each Fibonacci number is divided by its next larger neighbour. Thus, starting at the beginning with the first two ratios $F_1/F_2 = 1$, $F_2/F_3 = \frac{1}{2}$ (or 0.5 in decimals) and continuing along the sequence, we generate the successive numbers

$$1.000\ 000$$
$$0.500\ 000$$
$$0.666\ 666$$
$$0.600\ 000$$
$$0.625\ 000$$
$$0.615\ 385$$
$$0.619\ 048$$
$$0.617\ 647$$
$$0.618\ 182$$
$$0.617\ 978$$
$$0.618\ 056$$
$$0.618\ 026$$
$$0.618\ 037$$
$$0.618\ 033$$
$$0.618\ 034$$
$$0.618\ 034$$

which settle down to this strange value 0.618 034 ..., where the dots indicate the existence of more decimal places if we work to a greater accuracy than the six decimals given in the numbers above. In fact, in the limit of taking these 'Fibonacci ratios' on and on for ever, the number generated approaches closer and closer to $(\sqrt{5}-1)/2$ which, to the accuracy obtainable from my pocket calculator, is

$$0.618\ 033\ 989$$

but which, more exactly, is a number for which the decimal expansion continues endlessly without ever repeating. Such a number is called *irrational*, not for any reasons concerning lack of sanity (which you may be forgiven for suspecting), but in a mathematical sense concerning the fact that it can never be expressed as the *ratio* of any two whole numbers.

The Fibonacci family of numbers has been the subject of intense interest over the centuries for three separate reasons. The first involves the manner in which the smaller members of the sequence repeatedly turn up in the most unexpected places in nature relating to plants, insects, flowers and the like. The second is concerned with the significance of the limiting ratio 0.618 033 989 ..., often called the 'golden ratio', a number which seems to be the mathematical basis of everything from the shape of playing cards to Greek art and architecture. The third focuses on the fascinating properties of the numbers themselves, which find all sorts of unexpected uses in the theory of numbers. In fact, the literature on the Fibonacci numbers has now become so large that a special journal, *The Fibonacci Quarterly*, is devoted entirely to their properties, and produces several hundred pages of research on them each year as well as organizing occasional international conferences to boot.

Let us first look at the manner in which the smaller Fibonacci numbers appear in nature. You can, to begin with, nearly always find the Fibonacci numbers in the arrangement of leaves on the stem of a plant or on the twigs of a tree. If one leaf is selected as the starting point, and leaves are counted up or down the stem until one is reached that is exactly above or below the starting point (which may require going around the stem more than once), then the number of leaves recorded is different for different plants, shrubs and trees, but is nearly always a Fibonacci number. But what is more, the number of complete turns around the stem which need to be negotiated in the leaf counting ritual before the process begins to repeat itself, is also a Fibonacci number. Thus, for example, the beech tree has cycles of three leaves involving one complete turn, while the pussy willow has 13 leaves involving five turns.

In general, botany seems to be a veritable goldmine of Fibonacci numbers. Daisies are usually found with a Fibonacci number of petals so that, as one earlier commentator has put it, 'a successful outcome of "she loves me, she loves me not" is more likely to depend upon a knowledge of the statistics of the distribution of Fibonacci numbers than on chance or the intervention of Lady Luck'. Which Fibonacci number appears most frequently in the context of daisies I do not know (it is unfortunately winter as I write this; otherwise I would naturally do the necessary research myself) although reports of 21, 34, 55, and even 89 have been made.

Perhaps the most famous of all the appearances of Fibonacci numbers in nature is in association with the sunflower. In the head of a sunflower the seeds are found in small diamond-shaped pockets whose boundaries form spiral curves radiating out from the centre to the outside edge of the flower as shown in figure 1. If you count the number of clockwise and counterclockwise spirals in the pattern you will almost always be rewarded with consecutive numbers of the Fibonacci sequence. There are 13 clockwise and 21 anticlockwise spirals (count them!) in figure 1: these numbers are smaller than normally found in nature but they do make the picture drawing easier. Most real sunflower heads seem to have spirals of 34 and 55, although some smaller ones do have 21 and 34, while larger ones often contain 55 and 89; even examples with 89 and 144 spirals have been reported. But the sunflower is in no way special except that its seeds are particularly large and the spirals correspondingly easy to identify, and to count. The seed heads of most flowers, and many other plant forms such as the leaves of the head of a lettuce, the layers of an onion, and the scale patterns of pineapples and pine cones, all contain the Fibonacci spirals.

Some of the most careful studies have been carried out for cones on various types of pine trees. The spiral counts are most easily made when the cones are still closed; that is, fresh and green. Older open cones can be persuaded to close up again (in which state the spirals are much more easily seen) by soaking them in water. Further interesting questions may then be asked such as whether the cones are left-handed or right-handed. By this we mean does the higher-numbered spiral always go clockwise, anti-clockwise, or sometimes one and sometimes the other? The answer seems to be that overall there are about as many left-handed as right-handed cones, but that some trees are

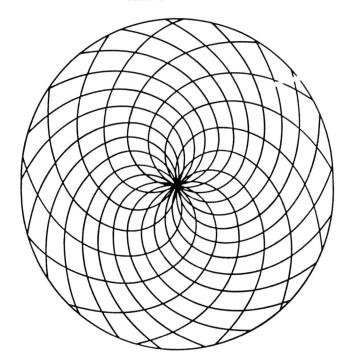

Figure 1

quite dominantly one or the other. What is particularly baffling is the fact that, even within a single particular species of pine, one tree can be dominantly left-handed while its neighbour is dominantly right-handed. The reason remains a complete mystery as far as I know. However, virtually all trees, no matter how left- or right-handed they may be, do produce some cones of each type.

Are any cones found which are not Fibonacci cones, you may ask? The answer is yes; but very few. Some, typically one or two per cent (and most often from a few specific species of pine), do possess 'maverick' cones. But even these are very often closely related to Fibonacci cones possibly having, for example, a double-Fibonacci spiral with a number pair like 10 and 6, rather than the more normal 5 and 3.

Since this chapter is entitled 'The Fibonacci Family and *Friends*', the time has now arrived to ask about the identity of some of these 'friends'. If we think back to how the Fibonacci numbers were made up, that is, from the equation

$$F_n = F_{n-1} + F_{n-2}$$

which calculates the nth number from its two smaller neighbours, it becomes apparent that the whole sequence is not completely determined until the first two numbers have been chosen. The Fibonacci series starts with $F_1 = 1$ and $F_2 = 1$, after which the above equation determines the rest. But there is really nothing obviously special about these two starting numbers and you could choose any other values for them and, using the same defining equation, derive a completely different sequence of numbers.

The most famous of these related families of numbers is the sequence of Lucas numbers, named after the French mathematician Edouard Lucas. It chooses the next simplest starting assumption with $F_1 = 1$ and $F_2 = 3$. Note that putting $F_1 = 1$ and $F_2 = 2$ (which looks simpler) merely repeats the Fibonacci series in a very slightly perturbed form and so does not present us with anything new. The Lucas numbers, on the other hand, are a quite different set from their Fibonacci relatives and begin as follows:

$$1, 3, 4, 7, 11, 18, 29, 47, 76, 123, 199, 322, \ldots .$$

They are usually represented by the symbols L_n, with the L in honour of Mr Lucas. Although it is obvious from what was said above that the Lucas sequence is only one of many such related sets, it is also of interest in the present context because these numbers sometimes show up in nature as well. For example, Lucas sunflowers have been reported. They are certainly rarer than their Fibonacci counterparts, but specimens with as many as 123 right-spirals and 76 left-spirals have been observed and carefully classified.

No-one really knows why these Fibonacci patterns, or less frequently Lucas patterns, appear in nature. In fact there are nearly as many proposed explanations as there are scientists willing to express an opinion. One of the less bizarre is a suggestion that the Fibonacci spiralling of leaves around a stem gives the most efficient exposure of the surfaces to sunlight. This possibility could actually be checked out mathematically, but I do not know whether anyone has yet gone to the trouble of performing such a calculation. Other less likely (and certainly less verifiable) explanations have involved some supposed preference of pollinating insects for 'numerical patterns' leading eventually, via an evolutionary process, to a dominance of Fibonacci geometries. In truth, your guess is probably about as good as anybody else's.

Let us now think a little more about that limiting ratio 0.618 033 989 ... , the so-called golden ratio, which is eventually generated by both the Fibonacci and the Lucas number sequences as they make their way steadily and laboriously out to infinity. Fascination with this particular number goes back for more than 2000 years. Although the 'ancients' probably did not understand its mathematical basis in the manner we have discussed, they certainly knew that art and architecture based on the golden ratio were unusually pleasing to the eye. They were therefore led to define the golden ratio in terms of geometry; specifically as the point which divides a straight line into two parts such that the ratio of the smaller to the larger is exactly equal to the ratio of the larger to the whole line.

For those of you who remember just a little of your school algebra we can now, by labelling the smaller part x and the larger part 1, write this geometric statement as

$$x/1 = 1/(1+x)$$

where the solidus (/) means 'divided by' and where $1+x$ is, of course, the length of the whole line. That this equation is indeed satisfied by the golden ratio can be checked directly by using your pocket calculator to verify that 1 divided by 1.618 033 989... is equal to 0.618 033 989.... But even better, if you can recall a little more school algebra, you can transform this statement into the quadratic form

$$x^2 + x - 1 = 0$$

and obtain the exact solution $x = (\sqrt{5} - 1)/2$.

If you draw a rectangle in which the ratio of the shorter to the longer side length is the golden ratio, then an extremely famous piece of artwork results known as the golden rectangle. The early Greeks referred to this even more reverently as the Divine Section. We show it in figure 2 and although at first sight it may not appear particularly worthy of such an accolade, it is in many ways a most remarkable construction. This is because over countless generations right up to the present day, most people see it as the most pleasing to the eye of all rectangles. As a result, a very large fraction of the thousands and thousands of rectangles which we meet in everyday life have dimensions which approximate those of the golden rectangle. Windows, parcels, book pages, photographs, match boxes, suitcases, playing cards, flags, writing pads, newspapers, and countless other examples all fall into this category. Without knowing why, the designers subconsciously prefer rectangular shapes close to that Divine Section. Why do they do it? Somehow the golden rectangle just 'looks right'; others are either too short and fat or too long and thin. For some reason not fully understood either by artists or psychologists the golden rectangle just has an aesthetic appeal.

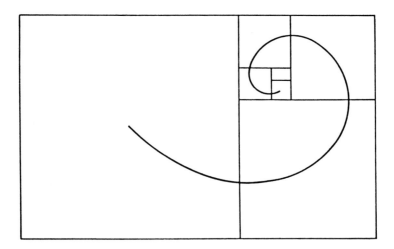

Figure 2

The golden ratio and Divine Section are frequently observed in Greek architecture and Greek pottery, as well as in sculpture, painting, furniture design and artistic design. It has been pointed out that the front of the Parthenon when it was intact would have fitted almost exactly into a golden rectangle. The golden ratio can also be found in the dimensions of some of the pyramids of Egypt, and Leonardo da Vinci became so fascinated by golden rectangles that he even co-authored a book about them.

Many of the great masters have proportioned their canvases with scrupulous regard to the golden ratio. Artistically two different kinds of geometric symmetry have been used which are related to the 'divine proportion'. One is the more obvious static relationship involving the number 0.618 03... via golden proportions or rectangles, but the other involves movement (or at least an imagined movement). The origin of this unlikely effect lies in another very special property possessed by the golden rectangle. It is that if this rectangle is divided into a square and a smaller rectangle, as shown in figure 2, then the smaller rectangle is also 'golden'. Moreover, continuing in this same vein, the smaller rectangle can also be divided into another square and another even smaller rectangle, and this rectangle too is golden. The process can obviously be continued in principle *ad infinitum*, creating an endless sequence of smaller and smaller squares and golden rectangles which spiral inward eventually to a point. Now, if we connect the corners or centres of these squares of ever decreasing size (or indeed of the golden rectangles, it makes no difference) by a smooth curve as shown in the figure, we generate a spiral popularly known as (yes, you guessed it) the golden spiral.

Looking at figure 2 with the eye of an artist we can now get a picture of 'whirling squares'. When incorporated into works of art in subtle forms this principle can be used to produce illusions of movement. The term 'dynamic symmetry' has been used to describe this and a number of artists, in particular the early 20th century American painter George Bellows, have made extensive use of illusions induced by whirling squares in their work. However, vestiges of the style can be traced way back to early Greek work.

The spiral generated by the whirling squares in figure 2 is not just any old spiral; in fact, it is very special and is the very same one which appears in the sunflower head of figure 1. Its proper mathematical name is the equiangular spiral or logarithmic spiral. It is 'logarithmic' because the algebraic equation which most simply defines it is written in terms of logarithms. But for those of you whose everyday life does not often bring you into contact with logarithms, the special nature of the golden spiral is much more simply grasped via its 'equiangular' property. This is the fact that any straight line drawn out from the centre of the spiral always crosses it at precisely the same angle as does any other such line; check this out using the figures.

Amazingly, it is this very special spiral which, for some reason, seems to be overwhelmingly favoured in nature. Shellfish, snails, most of nature's horns, tusks and claws, as well as all the Fibonacci-related cones and flowers discussed earlier, are nearly always found to be portions of equiangular spirals. Fibonacci's whirling squares generate a curve which, for some reason, nature finds particularly appealing. Even the great galaxies of outer space have arms of stars which whirl outward in gigantic equiangular spirals. Presumably, it is this subtle presence everywhere in nature of Fibonacci, his numbers, ratio and spiral, which makes these same proportions so pleasing in art. However, even quite apart from nature and art, the Fibonacci numbers and the golden ratio also have a purely mathematical fascination, and it is to some of these unlikely attributes that we now turn.

One of the more unimaginable concepts ever dreamed up by mathematicians is the so-called continued fraction. Everybody (at least everybody who has an inclination to pick up a book like this one) has a pretty good idea of what a fraction is. In particular, a rational fraction, which is the simplest kind, is just one whole number divided by another. Perhaps the simplest example of all is one-half, or $\frac{1}{2}$. In a valiant attempt to make the simple look more complicated we could rewrite it as

$$\frac{1}{1+1}.$$

But suppose that we got a little more ambitious and invented a fraction based on this simple form but which went on one step further in the fashion

$$\frac{1}{1+\dfrac{1}{1+1}}.$$

Those of you who still remember the rules for combining fractions will be able to evaluate the above and obtain the answer $\frac{2}{3}$. Similarly, if we proceed one step further in this same pattern to produce

$$\frac{1}{1+\dfrac{1}{1+\dfrac{1}{1+1}}}$$

those who know the rules can again find the simple rational fraction to which this peculiar object is equal; it is $\frac{3}{5}$. Continuing in the by now obvious pattern for yet one more 'storey' of fractional construction, we find that we have deduced a very complicated way of expressing the simple fraction $\frac{5}{8}$.

But now for the big step conceptually. What if this fraction went on and on in the same pattern forever

$$\cfrac{1}{1 + \cfrac{1}{1 + \cfrac{1}{1 + \cfrac{1}{1 + \cfrac{1}{1 + \cfrac{1}{\cdots}}}}}}$$

to become a 'continued fraction' in the sense of continuing *ad infinitum*? What would this number be? And how on earth could one ever calculate it? One clue is contained in the first four building-block fractions which we have already evaluated. They were equal to $\frac{1}{2}$, $\frac{2}{3}$, $\frac{3}{5}$, and $\frac{5}{8}$, and these are just the Fibonacci ratios F_2/F_3, F_3/F_4, F_4/F_5, and F_5/F_6. Is this mere coincidence or does it really imply that, when continued to more and more storeys, this odd looking fraction approaches closer and closer to the golden ratio 0.618 033 989 ...? In the infinite limit could the continued fraction set out above actually be equal to the golden ratio? And how can one get to infinity to find out?

Surprisingly, it is much easier to answer the question concerning the infinite limit than it is to calculate what number the 10th storey or 20th storey fraction is equal to. The secret is to look at the very top-most line of the infinite fraction and to think of it as '1 divided by (pause) 1 plus something'. Now we ask the question 'what is this something?' In general the answer is 'something pretty awful'. However, in the infinite limit, and only in this limit, the answer is easy; the 'something' is exactly the same infinitely continued fraction that we started with. If we call its value x then it is clear that this x must be equal to '1 divided by (pause) 1 plus x'. Written as an equation this looks like

$$x = 1/(1+x).$$

This is exactly the equation which we obtained earlier in the chapter when defining the golden section, and its solution $x = (\sqrt{5} - 1)/2$ is indeed the golden ratio. It follows that the simplest possible infinitely continued fraction (that is, one which is made up entirely of ones) is once again equal to that very special number 0.618 033 989.... That one has to continue the fraction all the way to infinity to get this is just another verification of the fact that the golden ratio is an irrational number.

Another mathematical quirk of the Fibonacci numbers is of particular interest since it was (to the best of my knowledge) first pointed out by Lewis Carroll, the creator of *Alice in Wonderland*. Lewis Carroll's interest in such things derived from the fact that he was, in real life, Charles L Dodgson, an

accomplished mathematician and, for most of the latter half of the 19th
century, mathematics tutor at Christ Church College, Oxford University. It
was there that he first met the real-life Alice who was one of the daughters of
the Dean of Christ Church; but that is another story. His Fibonacci puzzle was
usually presented as a geometrical paradox (or inconsistency) and used, in its
simplest form, as a method of 'proving' without words that 65 is equal to 64.

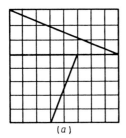

(a)

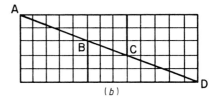

(b)

Figure 3 (a)   **Figure 3** (b)

Let us look at figure 3. Suppose that we first cut up a square of size 8 by 8 into
four pieces as shown in figure 3(a). These four pieces, consisting of two
identical triangles and two also-identical four-sided pieces, can then be taken
apart and reassembled in the manner shown in figure 3(b). In their new
arrangement the pieces form a rectangle but, and this is the puzzling part, the
side lengths of the new rectangle are 5 by 13. It therefore has an area of 5
times 13, or 65, compared with the original area of 8 times 8, or 64. What has
happened? A check of the side lengths of all the pieces which make up the
two areas reveals no obvious signs of cheating. Where then has the extra unit
of area come from in going from figure 3(a) to figure 3(b)? Think about it for a
little while before reading on. Can you see where the deception is?

Y ou will discover the secret if you draw the diagram exactly to scale on a
piece of graph paper, cut out the pieces according to the prescription of figure
3(a), and then try to reassemble them in the pattern of figure 3(b). For those of
you who have not made the effort, or whose unscientific everyday life does
not provide them with ready access to a piece of graph paper, let me explain.
It happens that, convincing though the figures are, the four pieces of figure
3(a) can in fact never be precisely fitted together to *exactly* make figure 3(b).
The points marked A, B, C and D on the figure should, if drawn accurately,
not be on a straight line (as they appear to be) but at the corners of a very
long and thin four-sided area (called a parallelogram) which makes up the
missing unit.

Very clever, you may say, but what has it got to do with Fibonacci? If you
look one more time at the figures you will notice that the integer side lengths
of the four component pieces are 3, 5 and 8, which are three consecutive
Fibonacci numbers. The important point is that there is nothing special about

the three particular numbers which we have used other than the fact that they are small and therefore easy to measure. We could have chosen any three consecutive 'Fibonaccis' and set up a similar paradox. If, for example, we start with a 13 by 13 square and cut it up in the same pattern as for the smaller example but using whole number side lengths of 5 and 8, then we can reassemble it to form an 8 by 21 rectangle to 'prove' without words that 13 times 13 is equal to 8 times 21 or, more explicitly, that 169 = 168. However, as we move to larger and larger Fibonacci numbers, the geometric 'misfit' which explains the paradox becomes more and more difficult to spot since the missing area of one 'unit' becomes an ever decreasing fraction of the whole picture. If, for example, we physically took a square piece of paper of side length 8.9 inches (which is quite large) and converted it in Lewis Carroll fashion to a rectangle of 5.5. by 14.4 inches (where 55, 89 and 144 are three consecutive Fibonacci numbers), then the greatest width of the 'slit' which makes up the misfit parallelogram is only about one-tenth of an inch and quite difficult to spot by eye.

Other kinds of mathematical fun can be had with Fibonaccis by arranging peculiar addition sums like this one below:

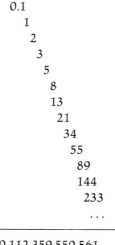

0.1
1
2
3
5
8
13
21
34
55
89
144
233
. . .

0.112 359 550 561 . . .

in which the Fibonacci numbers are added together, but writing each successive number so that it 'moves over' one digit to the right. Incredibly, this rather preposterous addition sum, when continued to infinity, has been proven to be exactly equal to the simple fraction $\frac{10}{89}$. Now my pocket calculator, which can only manage nine decimal places, assures me that $\frac{10}{89} = 0.112\,359\,551$, where it has obviously 'rounded off' the last decimal place that it can reach. Nevertheless, since $\frac{10}{89}$ is a rational fraction, and all such fractions are known to repeat their digit patterns sooner or later, the above decimals must eventually 'cycle', and so they do—after 44 decimal places. If the same addition pattern is set out using the Lucas numbers rather than the

Fibonaccis then another simple fraction $\frac{12}{89}$ is obtained. This too has a decimal form which repeats after 44 digits.

If similar addition sums are carried out moving each Fibonacci number over two digits to the right instead of one then we generate a decimal which begins as 0.010 102 030 508.... Its exact sum to infinity is also known and again is rational; $\frac{10}{9899}$ to be exact. In like fashion, moving over three digits at each stage generates yet another 'rational' namely $\frac{10}{998\,999}$. Simple fraction answers are also known for equivalent summations using Lucas numbers, as well as hosts of others which add and subtract the participating Fibonacci or Lucas numbers in an alternating fashion or which add only every other (or every third) sequence number and so on. Why should all these infinite sums lead to rational numbers i.e., to repeating decimal patterns? After all there are known to be infinitely many more (non-repeating decimal) irrationals than rationals among decimal numbers in general. I leave that to you to ponder at your leisure. Why should I be the only one with sleepless nights?

One other thought may now have crossed your mind. Earlier in the chapter I implied that, since the golden ratio was generated by an infinite continued fraction, it was irrational. With all these infinite sums leading to rational numbers some doubt may be in order. Fortunately, the irrational nature of the golden ratio is very easy to prove for anyone who ever got through the first chapter of his first algebra book. We merely start from its defining equation $x^2+x-1 = 0$ and assume that a rational solution p/q exists with p and q whole numbers. Reducing this fraction to its simplest possible form by dividing top and bottom by the same number whenever possible (e.g., $\frac{164}{64} = \frac{82}{32} = \frac{41}{16}$) we can always make sure that no integer (except 1) exactly divides both p and q. Writing $x = p/q$ now makes the defining equation look like

$$(p/q)^2 + (p/q) - 1 = 0.$$

Multiplying through by q^2 and rearranging the terms leads us to $p^2 + pq = p(p+q) = q^2$, or equivalently

$$(p+q) = q^2/p.$$

Since the left-hand side is a whole number, this equation says that p divides q^2 exactly (that is, without remainder). But if there is no integer which exactly divides both p and q (in mathematical language, if p and q have no 'common factor') then this is clearly impossible. It follows that the golden ratio x just cannot be written as p/q if p and q are whole numbers; in other words it is irrational.

Clearly, the Fibonacci numbers and their various offspring play a most extraordinary role in nature, art and mathematics. New mathematical extensions (such as the 'Tribonaccis'

$$1, 1, 2, 4, 7, 13, 24, 44, 81, 149, 274, 504, \ldots$$

in which the general term is made up by adding together the *three* preceding ones) seem to appear with every new issue of *The Fibonacci Quarterly*.

Applications also abound, although some interpretations with respect to art are arguably more romantic than reliable. But even if we allow for the fact that Fibonacci addicts will often twist almost any observation into some form of approximate relationship with these numbers (some have seen a golden spiral in the shoreline of Cape Cod while others have found approximate Fibonacci sequences in the sizes of insects on flowers, the distances of moons from their planets and in the radii of atoms in the Periodic Table of elements) the sum total of evidence is indeed persuasive. The Fibonacci sequence and the golden spiral are an important part of some recurring growth pattern; but the 'how' and the 'why' of it all remain a complete mystery. And all this from a theoretical family of 'abracadabric' rabbits conjured up in the mind of a 13th century lad who, it is recorded, was not exactly held in awe by his neighbours who referred to him disparagingly as Bigollone, 'the blockhead'.

3

Rising and Falling with the Hailstone Numbers

If anything at all is certain about the 'hailstone number' problem it is that its origin is shrouded in mystery. It certainly did not begin with an historic publication or even with any recorded exchange of letters between mathematicians insofar as I am aware. It does not appear to be of very great age, but seems to have turned up in a rather haphazard fashion at centres of learning all over the world during the last fifty years or so. Whether passed on by word of mouth or 'rediscovered' independently over and over again is not clear. In truth, the problem does not even have a name that is universally accepted. Some refer to it as the $3N+1$ problem, others as the Collatz problem (after a certain Lothar Collatz who, as a student in the 1930s, is credited by some as a possible originator). The description in terms of hailstone numbers is of recent vintage but, as we shall soon see, it does seem to offer a particularly apt visual perception of the entire phenomenon. In any event, nothing of significance was recorded in print about the problem until the 1950s. Since then, however, and particularly since 1970, it has become the focus of rapidly increasing attention. Prizes have been offered for its solution and a deluge of false proofs has unsuccessfully chased the prize money.

So what are hailstone numbers and why all the fuss? Perhaps, given their short and decidedly sketchy history, too much significance has been attached to them. On the other hand, they are unbelievably easy to define and yet they not only give rise to an unsolved problem, but one which (according to today's best mathematical minds) is likely to remain unsolved for many years to come. At least one such expert has been quoted as saying that 'mathematics is just not yet ready for such problems'. So let us give these numbers the benefit of the doubt and delve a little into their particular brand of mystery.

Hailstone numbers are produced in an extremely simple way by using the following rules. Think of a number; if it is odd, triple it and add one; if it is

20

even, halve it. Repeat this recipe over and over to each new number so obtained and see what happens as the progression continues. Let us investigate the very simplest cases by looking at the smallest possible starting numbers 1, 2, 3, and so on. Applying the rules we calculate the respective sequences:

$$1, 4, 2, 1, 4, 2, 1, 4, 2, \ldots$$
$$2, 1, 4, 2, 1, 4, 2, 1, 4, 2, \ldots$$
$$3, 10, 5, 16, 8, 4, 2, 1, 4, 2, \ldots$$

which all quickly enter the same 142142142 loop. Let us try again with a slightly larger starting number, say 7:

$$7, 22, 11, 34, 17, 52, 26, 13, 40, 20, 10, 5, 16, 8, 4, 2, 1, 4, 2, 1, \ldots .$$

This time it takes a little longer and the sequence gets up to a respectable high of 52, but then it crashes back down again and the final result is as before, an entry into the 142142142 endless loop.

The simple question to be answered is 'must all such sequences, regardless of starting number, eventually meet their demise in this same manner?' Although, as I have implied above, the answer to this question is not known at the time of writing, we can at least give a little bit of 'less than rigorous' consideration to the general situation. One might reason, for example, that since odd and even numbers occur with equal likelihood in the boundless sea of whole numbers, then one should at any point in the sequence be just as likely to be at an odd or an even value. Then, since the rules require that the odd number be more than tripled in going to the following step while the even number only gets reduced by a factor of two, the 'general' series (if by that we mean the usual situation, excepting some 'unlucky' specific cases) should increase forever. Could we then, perhaps, merely have been a bit unlucky in our first few specific examples above?

A little more testing is evidently called for. But, wait a moment, we do not have to try all the starting numbers in order. We can immediately see the demise of any number which has already appeared in any of the above sequences. Also, since any even starting number gets halved at the first step, we need not consider these either; some smaller odd number is bound to generate the same sequence (namely, the first odd number which appears in the even-number sequence). This reduces our continuing labour to the examination of the starting numbers 9, 15, 19, ... etc. These you can easily examine for yourselves and none survives long (all quickly crashing to the 142142 loop) until we reach starting number 27. And then, at last, we meet with a little adventure:

$$27, 82, 41, 124, 62, 31, 94, 47, 142, 71, 214,$$
$$107, 322, 161, 484, 242, 121, 364, 182, 91, 274,$$
$$137, 412, 206, 103, 310, 155, 466, 233, 700, 350,$$
$$175, 526, 263, 790, 395, 1186, 593, 1780, \ldots .$$

At this point in the sequence (the 39th number, or the 38th step of generation) we are safely past the 1000 barrier and seem to be going strong. If the sequence is followed still further, it gains even more in strength and finally reaches a 'high'of 9232 at the 77th step. But then disaster strikes and the 'crash' begins:

$$9232, 4616, 2308, 1154, 577, 1732, 866, 433, \dots,$$

until finally at the 111th step (112th number) we reach 1, the end of the line, or more precisely entry yet again into the 142142 loop. But this time, at least, we did get a ride for our money. The full saga of the trip is shown pictorially in figure 4.

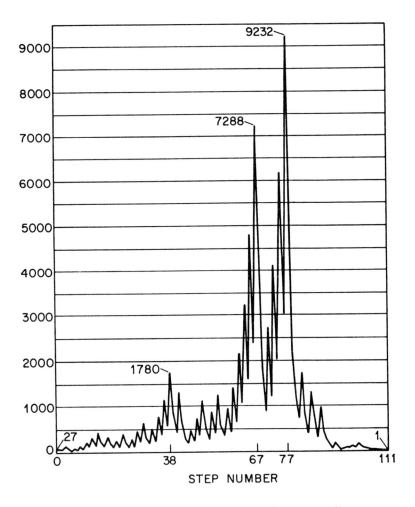

Figure 4 The sequence of hailstone numbers beginning with 27.

It was from pictures of this sort that the name 'hailstone numbers' was derived. The numbers rise and fall in a manner not at all unlike that of growing hailstones in a thundercloud—first caught in a strong up-draft and carried to great heights, then passing out of the supporting current of air to fall under their own weight, only to be caught up once more in an even more powerful draft to repeat the cycle. Within the thundercloud, however, the real hailstone is continuously growing in size so that eventually its fate is assured; sooner or later its weight exceeds the ability of even the most powerful up-draft available to support it and it plunges to earth. Must this same fate also inevitably await our hailstone number?

Clearly, something must have been wrong with our initial reasoning. Regardless of whether all hailstone numbers eventually fall 'back to earth' it now seems very clear that a great many of them certainly do. The weakness of the earlier argument can be seen by examining the sequence above for the starting number 27. Thus, whereas the number which follows an odd member of the sequence is always even by definition, the number which follows an even number need not necessarily be odd. Therefore, odd and even numbers do *not* have the equal probability of appearing in the series which the earlier argument assumed. Even numbers will inevitably outnumber their odd counterparts and therefore tend to offset the three-to-two advantage which the 'rules of the game' seemed at first glance to have given to the odd numbers. In fact, it is immediately apparent that only three sorts of nearest-neighbour pairs can occur in a hailstone sequence, namely, odd followed by even, even followed by odd, and even followed by even.

Suppose that we now ask 'what if the likelihood of these three combinations occurring at any point in the sequence is equal?' Well, since the first (odd followed by even) increases the hailstone number by a factor of three (the 'add one' part of the rule becomes negligible for general estimates involving primarily large numbers) while the second and third both decrease the hailstone number by a factor of two, the *average* result per step is a multiplication of 3 times $\frac{1}{2}$ times $\frac{1}{2}$, which is $\frac{3}{4}$. The suggestion now is that on average, when all the jaggedness is smoothed out of the hailstone curves, any large starting number, say one trillion or 10^{12}, would decrease by about 25% at each step and therefore inevitably finally reach 1 (in about 90 to 100 steps for the particular example cited).

Now I do not know how many steps are actually needed to reduce the hailstone series which starts with one trillion all the way down to 1 (although it apparently has been studied by computer and is known to fall finally into the 142142 loop) but, since it takes over 100 steps to settle the question for starting number 27, there must be enormous fluctuations in the number of steps required for similar starting numbers, even if the above reasoning is qualitatively correct on average. In fact, this reasoning still does not exclude the possibility of a fluctuation so large that for some starting number the hailstone may never come down. Now, although no computer will ever establish the existence of such a hailstone beyond doubt, since even the most powerful computers imaginable can only sample a finite number of steps, it is instructive to probe the hailstone sequences simply by generating them to as high a starting number as we can; that is, by 'number-crunching'. To start with, such a computer program is extremely easy to compose and, in addition, once it is running we do not have to do any more thinking for a while, but simply sit back and watch the hailstorms in action.

At the time of writing (1989) the most ambitious undertaking of this kind of which I am aware has been reported by the University of Tokyo. Apparently all numbers up to one trillion have already been tested and every single one eventually collapses to the 142142 loop. It certainly looks as if what goes up must indeed come down! And yet for many hailstones the 'trip through the thundercloud' is found to be quite eventful and some great heights are reached. Let us look at a few of the findings for the first 100 000 starting numbers.

Among the first 50 starting integers, 27 has the longest path back to unity. It is one of 111 steps (involving 112 numbers counting the starting number). As can be seen from figure 4 it sweeps up to the lofty height of 7288 at the 67th step, then falls dramatically to 911 before being caught in a new up-draft which carries it to even greater heights (specifically 9232 at step 77) before yet another plunge forces it finally, after a few lingering gasps, all the way down to 1 at the 111th step.

Beyond the first 50 starting integers the 'peak' at 9232 proves to be quite a barrier and is not surpassed until the starting number 255 is reached. This sequence is shorter than the one starting with 27, but rises dramatically up to 13 120 before coming rapidly back to earth. In fact, the peak at 9232 triggers the demise of all the longest sequences until we get all the way up to starting number 703 which continues for 170 steps and reaches a peak of 250 504. This starting number 703 is one of only two greater than 27 (and less than 100 000) which create both new records for length and height together. The other is 26 623 which continues for 307 steps and reaches a peak value of 106 358 020.

It seems clear from the above snippets of numerical information that the number-of-steps record increases rather slowly with increasing starting number. It is already 111 at starting number 27 and has reached only 350 at starting number 77 031 (which has the longest sequence for any starting number below 100 000). The peak values, on the other hand, increase much more dramatically, reaching the value 1 570 824 736 at starting number 77 671 (which is the highest peak for any starting number below 100 000). Think about this for a moment. It means that by this stage of our investigation the peak value to which the hailstone rises is more than 20 000 times its starting value. Moreover, this ratio of peak height to starting value appears to be growing rapidly with increasing starting number. Those up-drafts and sudden falls are quickly becoming more and more dramatic.

It has been suggested that for extremely large starting numbers N (with, say, a few hundred digits or more) the sequence *length* is reasonably 'well behaved', settling down, on average, to a value of about $24.64D - 101$, where D is the number of digits in the starting number N. This formula has no great theoretical foundation but has been based on many 'spot checks' of randomly chosen, extremely large starting numbers. For example, believe it or not, the sequence with starting number $N = 1 ...(998\ zeroes)...1$ has been followed by computer and descends to the 142142 cycle in 23 069 steps. With $D = 1000$ digits\the\formula would|have predicted 24 539 steps, an error of about 10%.

It is easy to see that the peak value reached by a hailstone number must always be even. It has also been proven that only an odd starting number can ever set a new peak record. In the case of starting numbers which set new sequence length records, however, there appears to be no theoretical restriction to odd or even. On the other hand, it does appear that most of the new length record holders are odd; the only exceptions below 100 000 are 6 (with sequence length 8), 18 (sequence length 20) and 54 (sequence length 112).

If a listing is made of *all* the sequence lengths and peak heights for the first (say) 100 starting numbers, a peculiar distribution is obtained—one which is definitely not random yet not easy to fathom. This you can easily do for yourself with the aid of a pocket calculator to hurry things along. One hundred starting numbers sounds like a lot until you realize that you do not have to do them all separately. Think back to the series for starting number 7 which we generated earlier:

7, 22, 11, 34, 17, 52, 26, 13, 40, 20, 10, 5, 16, 8, 4, 2, 1, ...

Not only does it tell us that the sequence length is 16 with a peak value of 52 but (since the second number is 22) it also tells us that the sequence length for 22 is 15 with a peak value of 52 and (since the third number is 11) that the sequence length for 11 is 14 with a peak value of 52, and so on. In fact this single 'hailstorm' gives us all the information we need for no less than 16 starting numbers. It also gives us a little bit of understanding as to why the same peak value arises for so many different starting numbers e.g., 52 is obviously the peak for not only the starting number 7, but also for 22, 11, 34 and 17. A supreme example of this is the peak value 9232 which appears in the sequence for 27. Since it occurs at step number 77, all 76 numbers before it must necessarily have this same peak value. In fact, of the first 1000 starting numbers, more than one third have this same peak value.

It is now clear why the distribution of peak values will be far from random; but what about the sequence lengths? Every possible length can certainly occur (this is very easy to prove by thinking of the starting numbers that are powers of two, namely 2, 4, 8, 16, 32, 64, ... which generate sequence lengths of 1, 2, 3, 4, 5, 6, ... etc) but, once again, some lengths appear far more often than others. Actually, they tend to form clusters and, in the year 1976, a string of no less than 52 consecutive starting numbers all with the same sequence length was published. What can it all mean? A smaller string, but with much smaller starting numbers, exists for the series from 386 through 391. This group is particularly interesting in that its members all have not only equal lengths but also equal peak values (of 9232 of course). Check them out for yourselves.

So where do we now stand concerning the 'hailstone conjecture'? Since every starting number up to 1 000 000 000 000 (that is, one trillion) is now known without doubt to fall back eventually to 1 and into the 142142 endless loop, it seems highly likely that all numbers do. Whether a simple proof exists or, if it does, whether it will be found in our lifetime, is uncertain. The problem is not really important enough to occupy the attention of serious research workers, although a great many mathematicians have given it more than a passing thought. So many, in fact, that at one time the current joke was that the problem was probably part of a foreign conspiracy to undermine serious mathematical research in the United States.

One thing seems particularly amazing to me as I ponder those one trillion sequences of numbers. It is that not a single one of them, no matter how long, contains the same number twice. This is truly astounding if you consider that with its 'hailstones' sweeping up and down, a sequence constantly passes through the same regions of number space. How do we know that a repeated number never occurs? Simply because if it did the pattern of numbers would repeat endlessly in a cycle (or loop if you prefer) and never fall finally down to 1, which we are informed that they all do. In a simple probability argument like the one which we used to persuade ourselves that the hailstone sequence fell on average by a factor $\frac{3}{4}$ at each step, such coincidences would be bound to occur eventually. The chances of getting through a trillion such sequences without finding a single coincidence of this kind would be unbelievably small. Think about it! Every time our sequence reaches an odd number after falling through a series of consecutive even ones, it starts to move back into a number territory where it has been before. This it does over and over again in almost all of the trillion hailstone sequences which have so far been checked through, and yet not a single repeated number has been found.

It is almost unthinkable that we could have been that lucky (or unlucky depending on your point of view) by sheer chance. We are driven to the conclusion that the numbers generated in the hailstone sequences are (appearances notwithstanding) far from random. They must have imbedded in them some precisely determined mathematical restrictions, one of which we have presumably stumbled upon. But if they are not random-like, then any probability argument concerning them is doomed. This immediately spells the demise of our $\frac{3}{4}$ argument, the only one we possessed which pointed to the inability of any hailstone number to 'fly forever'. Maybe, therefore, we should maintain an element of doubt. The book is not closed. Perhaps some hailstone numbers really can fly forever to higher and higher values in a boundless fashion—and, if not, at least get hung up in a loop a little more distinguished than the 142142 terminus. Since it has been claimed that there are no other cycles with a period of less than 400 000, such a loop, should it exist, would be impressive indeed.

All of this is quite fun to read about, you may be thinking, but what can I do to contribute, armed only with a modest pocket calculator (if that)? For the actual hailstone problem set out above the answer is evidently little, since powerful computers have already been set loose upon it. On the other hand, all hailstorms are not the same, they can come in many guises. In fact there are countless numbers of them, on most of which 'the hand of man has never set foot' (to use one of my favourite mixed metaphors). For example, instead of multiplying an odd number by 3 and adding 1, you could multiply by 3 and add absolutely any designated odd number, say 3, 5, 7, or even larger. For all these 'hailstorms' our earlier $\frac{3}{4}$ argument (to the extent that it retains any credibility among you at all) remains intact to suggest that no sequences of this kind can go on increasing forever. But now, at least in some instances, other loops *can* be generated.

Consider, for example, the hailstorm with an odd-number rule of 'times 3 and add 7' and the usual even-number rule of 'divide by 2'. The sequence generated by starting number 1 is

$$1, 10, 5, 22, 11, 40, 20, 10, 5, 22, 11, \ldots$$

where we immediately get caught in an endless 5, 22, 11, 40, 20, 10, 5, loop. Starting number 7 finds another loop in the form

$$7, 28, 14, 7, 28, 14, 7, \ldots$$

while starting number 9 comes down to 1 after a 25-step ride as follows:

$$9, 34, 17, 58, 29, 94, 47, 148, 74, 37, 118, 59, 184, 92, 46, 23, 76, 38, 19, 64,$$
$$32, 16, 8, 4, 2, 1, \ldots.$$

This sequence, as we can see, crashes because it 'hits' a power of two (namely 64, which is 2^6) which drops it all the way to 1 'like a stone'. Interestingly, this suggests another (shaky) probability argument. It might be claimed that eventually (since the powers of two are infinite in number) any hailstone number sequence must, if it goes on long enough, be certain to alight on one of them if it is not cycling, and thus come tumbling down to earth. In fact this argument, to the extent that it is worth anything, can be used equally well for other hailstone types in which we multiply odd numbers by 5, or 7, or 9, and add (say) 1, again (as always) dividing even numbers by 2. For the latter sequences the old $\frac{3}{4}$ argument no longer applies (check it out) but moves to a $\frac{5}{4}, \frac{7}{4}$ or $\frac{9}{4}$ argument which predicts that (on average) each number will now be larger than its predecessor by a factor of $\frac{5}{4}, \frac{7}{4}$ or $\frac{9}{4}$, etc. The implication is that these new hailstone sequences will, unless they are unlucky, go on forever, getting larger and larger without bound.

Here, therefore, we have a particularly interesting situation. Our two probability arguments are clearly in conflict. For example, in the 'hailstorm' for which we multiply odd numbers by 5 and add 1, one argument says that each term, on average, should be $\frac{5}{4}$ times as large as the one before it and that the sequence should consequently grow forever; the other argument concerning the powers of two says that any such sequence which has pretensions of tending to infinity is (in spite of the $\frac{5}{4}$ rule) bound to be 'unlucky' and never make it. Which do we believe?

To my knowledge no vast amount of research has been performed on this hailstorm, so that the road is open for your own efforts. I shall accompany you only a small way. Thus, for starting number 1 we find

$$1, 6, 3, 16, 8, 4, 2, 1.$$

The power of two argument soon won that one! Since starting numbers 2, 3 and 4 are already included in the above series, they also crash to a final value of 1 (or to a 1, 6, 3, 16, 8, 4, 2, 1, ... loop if you prefer). The next starting number of interest is therefore 5. Following its sequence we find

$$5, 26, 13, 66, 33, 166, 83, 416, 208, 104, 52, 26, 13, \ldots$$

and generate a 'non-trivial' loop (by which we mean a loop which does not contain the number 1) running from 13 up to 416 and back again to 13. As for starting number 7 I will tell you only that it is quite an adventure. Go ahead and investigate. Try some other starting numbers. Then try any of the other almost limitless kinds of hailstorms and perhaps uncover some 'conjectures' of your own. Unless, of course, you fear becoming part of that international conspiracy to undermine the study of 'serious' mathematics in this, or any other, country.

4

Lies, Damned Lies, and Statistics

It was the British Prime Minister Benjamin Disraeli who once warned his parliamentary colleagues that 'there are lies, there are damned lies, and there are statistics'. Just how the science of statistics has gained such a poor reputation is not clear, but the implication is that the average 'layperson', intelligent or otherwise, knows so little about this science that he or she can easily be persuaded to accept completely unfounded conclusions by the devious or merely by the innocently ill-informed. The major problem is that in matters of probability common sense and intuition are often very poor guides. Indeed, there are a great many simple illustrations of this fact so that it is, perhaps, a good idea to start with a little example; one which might well confront any good citizen in the execution of his duties as a jury member.

The case in question involves a hit-and-run accident to which there was an eyewitness. This witness reported that the vehicle involved was a taxi and that its colour was blue. Now it so happens that in the city in which the accident occurred there are only two taxi companies, one which operates green vehicles and one which operates blue ones. The case seems clear enough; in all probability the hit-and-run driver should be sought among the drivers of the 'blue-cab' company (excluding for the moment the unlikely possibility that the vehicle in question had been stolen).

But wait a minute, there is a slight complication. It was getting dusk when the accident happened and, under test conditions, the witness was only able to correctly identify the colour of a green or blue vehicle 80% of the time. Well, you may say, that does make the identification a little less certain but still (and surely without question) the most probable situation remains that it was a blue taxi which was involved in the incident. What other conclusion could a responsible juror possibly come to? This piece of evidence should therefore take its place alongside all the additional information which is presented in helping to come to a fair verdict. Do any of you quarrel with that assessment? I doubt it very much and yet, for this particular event (as we shall set out below), it is completely unfounded.

The reason is that a vitally important question, without which the information so far gleaned is quite useless, has not yet been asked. It is 'what percentage of the taxi-cabs in this particular city belong respectively to the green-cab and blue-cab companies?'. It turns out that the answer to this question for our case is that the green-cab company is much larger than the blue-cab company and operates no less than 85% of the taxi-cabs in the city. If the witness incorrectly determines the taxi-cab colour 20% of the time then the whole picture now changes. Let us see why.

Without getting into any details of complicated mathematics or probability formulas we can quickly get into the spirit of statistics simply by imagining that the witness observed many (say 100) such accidents, instead of just one, and reported observations on all of them as regards the colour of the vehicle involved. By the laws of probability, about 85 of these accidents might be expected to involve green taxis and about 15 of them blue taxis. Of the 85 green ones the witness would incorrectly identify about 20%, or 17, as blue. And of the 15 accidents involving blue taxi-cabs he will correctly identify about 80%, or 12, as blue. Thus, of the 29 times that the witness reports seeing a blue taxi-cab involved in an accident he is wrong no less than 17 times—an error rate of almost 60%. The report of the witness that the taxi-cab involved was blue is therefore most probably a misidentification of a green taxi-cab.

The correct conclusion to be drawn from the evidence presented is therefore that the taxi involved was most likely a green one—although the odds are now so close to even that it would perhaps be best for the juror to disregard this particular piece of evidence altogether. We can now begin to appreciate Disraeli's hearty scepticism of statistics. Many of the conclusions of the theory of probability do indeed seem to run counter to the dictates of 'good old common sense'.

This troubling tendency of statistical theory, in spite of being mathematically quite sound, to present the mind with unbelievable conclusions goes way back through the centuries. One of the most fascinating of the earlier examples was first discussed in the 1730s by the Swiss mathematician Daniel Bernoulli and involved a game of chance. The idea was to toss a coin until it came down 'tails'. If this happened on the very first toss then it would be designated as worth a 'win' of 2^1 (that is 2 to the first power, or 2) tokens, with a token being valued at whatever you choose, say a dollar in modern money. If the tail did not appear until the second toss then the 'win' would be designated as 2^2 (or 4) dollars. If it did not occur until the third toss the win would be worth 2^3 (or 8) dollars, and so on. The question to be answered was 'what would be a fair amount of money for a coin tosser to pay the 'bank' for the privilege of playing this game?' By 'fair', of course, we mean that if the game were to be played over and over an endless number of times then the bank and player would 'come out even'.

Now mathematically speaking this is not a difficult question to answer. The problem is that the answer is difficult, if not impossible, to believe. We reason as follows. Since there is one chance in two of tossing a tail at the first try, there is one chance in two of winning 2 dollars at the first throw. If this were the *only* way of winning then a 'fair' price to pay for playing would obviously be 1 dollar. However, one can also argue that there is one chance in four that the game will finish after two tosses and therefore net you 4 dollars. Again, if the game 'payed off' only for this particular situation, then it is again clear that a 'fair' price to pay would be 1 dollar. In fact, the argument is quite general; if we ask 'what would be a fair payment for the privilege of playing the game in which I win *only* if a tail appears first on the nth toss (with n being any integer I choose)?' then the answer is always just 1 dollar. But the real game entitles me to win no matter when it finishes; that is, on *all* values of n. The 'fair' entry fee must therefore be the sum (or addition) of all these 1 dollar contributions for every possible end point; i.e.,

$$1 + 1 + 1 + 1 + 1 + 1 + 1 + \ldots$$

where the dots symbolize continuing forever. But this is an infinite amount and implies that no amount of money in the entire world would be sufficient to make this a fair game. In other words the game would favour the player even if he was required to pay a billion dollars for the privilege of playing. But would you be willing to offer up your entire wealth (be it sadly less than a 'billion') to play this game just one time? The mathematics says that you should jump at the chance. What has gone wrong?

One problem centres on the fact that the 'bank', in real-life games, has only a finite amount of money to pay out in possible winnings. Suppose, for example, that our particular banker could pay out wins up to 2^{20} (or a little over one million) dollars. All games which took more than 20 tosses to get a tail would then have to be counted null and void. With this restriction, what now would be a fair amount to pay for the privilege of 'having a gamble'? Using the earlier arguments we now only have to add 20 ones together in the previously infinite addition sum to obtain the answer, namely, 20 dollars. With this bet you would lose if a tail came up on any of the first four tosses, but win if it did not appear until the fifth toss or later—and the win might possibly be as much as a million dollars. Now the game does not seem quite such a bad proposition, does it? Has our intuition failed us again? Maybe, but that infinite limit still seems a bit hard to swallow. And so it was back in Bernoulli's day. This was a time before a sound mathematical grasp of the infinite was available and the result was the object of a great deal of controversy.

There is, moreover, an additional complexity. It concerns the fact that in real life the brain does not find an infinitesimal chance of winning an infinite amount of money a very tempting proposition, regardless of what mathematics has to say about the odds involved. And this is particularly true if the privilege of playing is costly. Let us, for example, consider a lottery in which there is one chance in a million of winning one million dollars. The 'fair' price of participation would be 1 dollar and (as the State Lottery system in America, which offers far poorer odds, has so convincingly demonstrated) most people are more than happy to participate. But now suppose that there was one chance in a million of winning a billion (10^9) dollars. The fair price for one ticket would now be 1000 dollars. This lottery is just as 'fair' from the point of view of mathematical odds as was the former, but now the almost certain (999 999 out of 1 000 000) loss of a very substantial amount of money (1000 dollars) would deter a far greater proportion of potential players from taking the risk. This effect becomes stronger as the amounts are raised. Who, for example, would risk 100 000 dollars (even if he could borrow it) on a one-in-a-million chance of winning 100 billion dollars (or even a trillion dollars), even though the last bet would be a fantastically good one from the sole point of view of strict mathematical odds?

The brain, you see, interprets things quite differently. The almost certain loss of a lifetime's potential earnings is far too great a price to pay for the remote chance of winning an amount of money so large that one could not conceivably spend more than a tiny fraction of it, even by indulging in every luxury and vice imaginable. Real-life goals are simply not determined by the favourability, or otherwise, of mathematical odds alone. Unless inflation really gets away from us, winning 10^{90} dollars is, from a practical point of view, no different from winning 10^9 dollars.

At the opposite end of the monetary scale, people are almost always ready to risk a trifling amount on a gamble even if the odds are absurdly adverse. The reasoning presumably goes something like 'we shall never miss the small amount needed to enter and, after all, somebody has to win, and we stand as good a chance as anyone else!' For example, if you are on the average 'junk-mail' list you may well, over the years, have received several notices of the kind which announce that 'you may already have won $100 000'. The implication is that some company has allocated a number to each family so notified and has already selected the winning number, which just might be the one in your envelope. However, you can collect only if you reply. Is it really worth answering these?

In the United States the larger operations involved in these kinds of 'giveaways' may easily mail such offerings to a substantial fraction of all the homes in the country—10 million could well be a conservative estimate. If one accepts this 10 million figure, then your chances of winning are one in 10 million. It follows that you can expect to win, on average, $100 000 divided by 10 million, which works out to be just 1 cent! This is far less than the cost of the postage stamp necessary to return the 'blurb' and enter the game. Cool-headed statistics therefore shout aloud that (on average) you can expect to lose more than 20 cents each time you play and that consequently only a fool would enter such a scheme. But try telling that to the winner! In fact such giveaways are joyfully entered by a substantial fraction of the persons solicited, who remain convinced that, as something for nothing, it is the best of all possible wagers.

The above scenario is one for which practicality overrides mathematical pronouncements for perfectly logical reasons. For some people the relief from the dull repetition of everyday life which is afforded by the excitement of an (usually short lived) anticipation of $100 000 may well be worth the price of a postage stamp in some very real sense. On the other hand, many other cases are readily found for which common sense reaches erroneous conclusions in matters of statistics with no redeeming features whatsoever. Perhaps the most common of all such misconceptions involves what the statistician loftily refers to as 'the principle of regression to the mean'. This principle was initially the notion of Sir Francis Galton, an English gentleman-scientist of the 19th century. It says quite simply that in any series of purely random events clustered about some average value, an extraordinary event is (just by the 'luck of the draw') most likely to be followed by a more ordinary event. Thus, it has been pointed out, very tall fathers tend to have slightly shorter sons on average, and very short fathers somewhat taller sons. If this were not so there would by now be a large number of 100 feet tall and 1 foot tall gentlemen walking about.

In the United States, particularly among sports fans, the strongest manifestation of this is the so-called 'sophomore jinx'. It refers to the fact that a new sports hero or heroine who bursts upon the scene with an unbelievably impressive first season very often has, by comparison, a disappointing second (or sophomore) year. In order to be more specific let us express the phenomenon in baseball terms. The choice is not exactly an arbitrary one since baseball, being so swamped with statistical data, is an ideal breeding ground for misconceptions of this kind. Consider a pool of equally talented pitchers who on average would be expected to win, say, 60% of their games in a season. Statistics, by their very nature, will (of course) not allow this ideal conformity to take place in a real season. Some of these pitchers will do better and others worse than their talent-determined 60% win percentage would justify. In particular, one or two will post win percentages well above 60%, possibly as high as 80%, purely as a virtually unavoidable consequence of perfectly normal statistical fluctuations.

But what will the press and the fans conclude? Will they make any reference to statistical fluctuations? Of course not! These few sporting fortunates (fortunate at least for one season) will be hailed as super-heroes. Very occasionally the designation may be justified but usually, come the second season, the principle of regression to the mean exerts itself. As a result there is an overwhelming probability that these same 'super heroes' will now post significantly worse records. The sophomore jinx surfaces and inevitably enters the vocabulary of the sports commentator. The cause will naturally be attributed to every imagined pressure or post-super-hero overindulgence in the good life. The overwhelmingly most likely explanation of 'regression to the mean' would have little journalistic appeal, even if it were recognized. It has never, to my knowledge, made the sports pages.

This very same 'regression' principle is also frequently misinterpreted to imply that criticism is a much more efficient inducement to progress than is praise. Let us see how this comes about. Imagine, for example, a number of pupils engaged in any endeavour whatsoever. With effort they will presumably increase their skills as the days and the weeks go by but (again because of the dreaded statistical fluctuations) this improvement will not take place in a manner which is predictable and smooth. In any particular test some, just by sheer good (or bad) fortune, will perform above (or below) their capability, where by capability we mean the level at which they would perform if averaged over a large number of such tests. Those who overachieve on the test will doubtless receive the lion's share of the praise and, by our now familiar statistical law of regression to the mean, are likely to revert closer to their (lower) true capability in a subsequent test. Exactly the reverse is true for those who underachieve on the first test for statistical reasons and are thereby subjected to criticism. All the evidence, to those unfamiliar with the true nature of statistical fluctuations, therefore points to the value of criticism over praise in inducing an improvement in performance. Possibly Disraeli, to his credit, would have been more sceptical.

A closely related sporting misconception is that of 'momentum' or 'the hot streak'. Although such concepts are not necessarily entirely without substance, it is overwhelmingly likely that the events which are interpreted as examples of this phenomenon are nothing more than randomness in action. Let us take a look at a typical season of play in a sport where (for simplicity) we assume that tied games are not allowed (baseball and basketball are two good examples). Let us also for simplicity assume that all teams in a particular league are of exactly equal ability. This means that the result of any particular game has the same probability of going to team A or team B as a coin toss has of coming down heads or tails. Statistically, what can we expect to take place during a season's play? Will all teams finish up with exactly equal won–lost records? Of course not! In fact it may be of some interest to you to play out a season of this kind, using a coin flip to decide matches. At the end of the 'season' there will be a most successful team and a least successful one, and sports writers would no doubt attribute the success of the one and the failure of the other to everything under the sun except statistics. Our point is not that the differences of talent between teams play no important role, but that the most talented team will often fail to win the 'championship', not through any lack of determination or imagined lack of character, but simply because of the inevitable role played by statistics.

Within a season, involving (say) 100 games for each team, other dramatic effects can be expected in our coin toss league of teams with identical abilities. Strings of games will be won consecutively or lost consecutively. This is the 'momentum' effect. If, for example, your last game was lost, then there is statistically one chance in 2^n that your next n games will also be lost due to no particular fault on your part. With $n = 6$, this leads to a probability of one in 64 that a string of seven games will either be won or lost consecutively *by sheer chance*, and such a string is quite likely to occur in a 100 game season. If these are losses then the team will be 'in a slump' and concerned about how to 'snap out of it'. On the other hand, if they are wins, then the team will have 'momentum going for it' and possibly even be credited with having discovered some 'winning formula'. Inevitably, however, the law of regression takes over and these moments of depression or elation pass.

It is, of course, possible to establish whether any effects other than statistical fluctuations are playing a role, but sports analysts rarely do, and even the players often come to completely false conclusions when any effort is made to establish some credibility for 'momentum'. One very informative piece of statistical research was recently carried out on professional basketball players. Both players and fans tend to believe that players shoot in streaks; that during a game a player has a time when he is 'hot' and every shot goes in. So ingrained is this belief that team members will actually try to get the ball to the player with the 'hot hand' in order to cash in on this effect.

But does the effect really exist? Research workers at Cornell University studied the detailed records of about 50 games of a professional basketball team. Although the players themselves thought that they were about 25% more likely to make a basket after a hit than after a miss, the researchers found that the opposite was true. In fact, a player was up to 6% more likely to make a basket after a miss than after a hit. Thus, in this case all those hot and cold streaks which inevitably occur due to statistical fluctuations were, in fact, slightly less impressive than if chance alone were responsible. Momentum was not only absent, but its imagined presence was quite probably actually hurting the team's overall performance. The most likely explanation is that a player who has (by chance) made a few consecutive hits, thinks of himself as 'hot' and either attempts a more difficult shot than he otherwise would or is more carefully guarded by defenders. The player who has missed a few, on the other hand, searches for a safer shot to restore his confidence, or is perhaps less closely pursued by defenders.

Whether or not a similar situation (in terms of overconfidence) exists in the winning and losing streaks of teams has not, to my knowledge, been studied. But regardless of this, nothing is going to deter the sportswriter from searching for an explanation for every conceivable statistical fluctuation. Long exposure to chance processes by no means guarantees that any more people recognize them as such. There is no sales potential for the sportswriter in discussing the subtleties of statistics, even if the reader (or writer) were able to appreciate them.

Quite generally it is very easy to demonstrate the complete abandonment of logic by the majority of people by asking questions concerning the making of choices based on the probable outcome of certain events. In this context, mathematically identical situations can induce unbelievably different responses depending solely on the manner in which the question is posed. Thus, for example, research workers have found that when people are offered the choice of a sure gain of $3200 or an 80% chance of winning $4000 and a 20% chance of winning nothing, they overwhelmingly opt for the sure gain

of $3200. If the situation is reversed and the choice is between a sure loss of
$3200 or an 80% chance of losing $4000 and a 20% chance of avoiding loss
altogether, the response is almost exactly the opposite with up to 90% of
respondents now opting for the gamble, thereby hoping to avoid the loss.
From a purely mathematical standpoint there should be no preference for
either option in either case. But apparently, in the real world, statistics count
for little and a dramatic pattern emerges; people tend to avoid risks when
pursuing gains, but will accept almost any risk to avoid taking a loss
regardless of what the mathematics dictates.

Perhaps the kinds of statistics which have lent most credibility to Disraeli's
original quotation are those which contain so-called 'confounding factors'.
These are hidden complications (usually unrecognized and often completely
unknown) which actually make any mathematically sound conclusions impossible to draw. They can most often be found in what we think of as
sampling errors. How, for example, can we be sure that two groups of people
who are about to be compared in some way are 'random' as regards other
properties which may (in a manner completely unknown to us) dramatically
influence the results? The often disastrous consequences of ignoring the
dreaded 'confounding factors' are easily illustrated by giving an example or
two.

Consider first the matters of living and dying in two fictitious towns of
Worksville and Snoozeville, each with a population of about 100 000. In a
comparison of mortality rates, it is found that Snoozeville has about 1500
deaths per year while Worksville's undertakers are less active in dealing with
only about 1000 deaths annually. The conclusion which so obviously
suggests itself is that Worksville is (for some reason) a healthier place to dwell
than Snoozeville. How can it be otherwise? Statistics do not lie! But then
some slightly more diligent statistician notes a peculiar thing. In spite of the
above numbers, it turns out that for every single age group (that is children
0–10 years, teenagers, twenties, etc, all the way up to the eldest citizens in the
'over 100 years' category) there are more deaths per year in Worksville than
in Snoozeville. This (equally obviously?) establishes exactly the opposite
conclusion, and from the very same set of data.

Has our computer made an error? Has somebody mixed up the figures?
Apparently not; careful rechecking shows that all the data are correct. The
problem can therefore only reside in our interpretations of the findings. One
of them is quite evidently false. The clue has been provided by the names of
the towns. Worksville is primarily a town with lots of new job opportunities
and is consequently populated by many young families. Snoozeville, on the
other hand, is largely a retirement community with little in the way of
employment for the younger set. Not surprisingly, therefore, the confounding
factor here is age. The age distribution in the two towns is completely
different. With most of its inhabitants in their retirement years it is hardly
surprising that the total mortality rate should be greater in Snoozeville—and
this would be expected regardless of the local health environment. The

essential point is that, without some knowledge of the respective age distributions, no conclusions whatsoever can be drawn from the gross mortality figures regarding the environmental health factor. The correct interpretation (if there are no more unrecognized confounding factors around) is that deduced by including age; i.e., in spite of the overall death rate, Snoozeville does appear to be a healthier place to live than Worksville.

A more ludicrous (but perhaps more amusing) example of the 'age' confounding factor has been given in the story of an enthusiastic amateur statistician who, comparing scholastic performance with every conceivable variable, finally noted an impressive correlation with shoe size. After giving a standardized test to all students at a series of elementary schools he was confronted with the inescapable fact that pupils with the largest feet consistently obtained the highest scores. Only later, to his great embarrassment, did it finally dawn on him that shoe size also correlated rather well with age. The correct conclusion was therefore the remarkably unsurprising one that older children tended to outperform the younger ones both in accumulated knowledge and in shoe size.

The general problem indicated by these rather obvious examples is nonetheless a very serious one. It is probably the statistician's most difficult task of all to assure himself or herself that no unrecognized confounding factor lies hidden in the sampling groups which are being tested. In fact, it is generally impossible to be *absolutely* certain that such a factor does not exist. A very good example of this involves the early work done on lung cancer mortality rates before the role played by cigarette smoking was suspected. Many completely false statistical correlations and conclusions were recorded in those years due to the significantly different proportion of smokers in the groups under study. At the time, smoking was completely unsuspected as possibly influencing the results in any way. With hindsight we now judge all of these early studies to be biased, and reject them all as completely unfounded due to the 'smoking confounding factor'.

A quite different, but equally disastrous, situation can arise when persons without statistical expertise genuinely misinterpret statistical events. This usually occurs when the errors made are systematic rather than random. In such a case, for example, the several members of a committee may all (in the absence of proper statistical counsel by experts) come to the very same unjustified conclusions from the data and hence not feel any need for expert opinion. In a political context this can often result in tax money less than well spent. One example (happily of no political consequence) is often cited in the form of the question 'how many persons must be in a room before there is a greater than even chance of at least one shared birthday?' The correct answer is 23, and it is claimed that the overwhelming majority of non-statisticians in the population-at-large find this to be an unbelievably small number—guesses in the 50 to 70 range being most common.

The implications of systematic subjective errors of this kind can sometimes have serious implications for public policy. One which has surfaced in recent years concerns the incidence of rare diseases in small communities. The problem has two separate facets. Firstly, one cannot expect that small numbers of cases of rare diseases will be equally distributed among towns and villages of comparable populations even if the chances of contracting them are exactly equal in the various locations. Natural statistical fluctuations, governed by a law known as the 'Poisson distribution', produces clusters of events in some locations and none in others. Thus, for example, if there are an average of two deaths per community from a particular disease per year, then the Poisson distribution tells us (in any particular year) to expect about 13% of the communities to have no deaths at all, and about 5% to have as many as five or six deaths, and this for statistical reasons alone. But confusing the issue further is the fact that, if statistics are gathered for more than one rare disease, then the shared-birthday misconception may also intrude to make a correct assessment of the situation even less likely. By this we mean that a community with a seemingly high incidence of more than one related disease may also arise by sheer chance much more often than the 'intelligent layperson' would be likely to expect. The situation is fraught with danger since, when people are randomly mistaken in their assessment of risk, then the errors tend to cancel out and no great harm is done. If, on the other hand, as in the shared-birthday and Poisson-distribution phenomena, the errors are systematically biased to an overwhelming degree, then they can lead to a serious waste of scarce resources in a misguided effort to protect us all from imagined hazards of all kinds.

Finally, before passing on to a new chapter, some mention should perhaps be made of the most sensationalized misuse of statistics, sometimes referred to as the hidden message phenomenon. It generally goes like this: in some ancient structure extraterrestrials have left, for us of the future generations, some secret message which is just asking to be decoded. The Great Pyramid of Giza has often been used as a favourite hunting ground for 'messages' of this kind. The idea is to measure every conceivable length and angle to be found, and then to correlate them with some other more noteworthy numbers like pi, or the earth's diameter, or the distance to the moon (or the sun or the stars, it does not particularly matter) measured in whatever units are most favourable. Shifting the decimal point, of course, is always allowed since this not only enables us to better appreciate the inscrutable nature of the aliens, but also gives us a much better chance of producing startling 'coincidences' from perfectly random events.

After finding such a coincidence the reasoning then proceeds with statements like 'the ratio of this column height to that diagonal room dimension is almost exactly the square root of pi. The Egyptians of the Giza building period could not have known pi to that kind of accuracy. Where could they possibly have got the information from?' Here comes the extraterrestrial implication, and the sensationalism follows from an eager press without too much encouragement.

The secret of success in this misuse of statistics is first to generate a veritable garden of 'significant' numbers. You can produce thousands of them just by looking at the solar system; planetary dimensions, masses, densities, and every conceivable ratio between them. Mix in an occasional multiplication by pi or one of its simpler powers or roots and you are all set to go. Because the rules of the hidden message game allow you to play around with the decimal point (finding something 100 times bigger or smaller than a particular significant number is, after all, really just as impressive as finding something actually equal to the number), obtaining a truly impressive coincidence even amongst the dimensions within your own house is almost guaranteed. Not that the press would be easily convinced of extraterrestrial involvement in the building of your house. The final secret (from a publicity point of view) is therefore to go off and find yourself a suitably ancient monument, the more ancient the more impressive. Good luck!

The reason that these number coincidences so often convince the layperson of the presence of something supernatural is that the entire list of thousands of 'significant numbers' is, of course, never revealed. Attention is entirely focused on that single coincidence which has been discovered, as if it were virtually the only candidate available. The possibilities are endless. For example, if you divide the height of the Sears Tower in Chicago (which is the world's tallest building) by the height of the Woolworth building in New York (which was also the world's tallest building when completed in 1913) you get the result 1.836. This is precisely one thousandth of what you also get if you divide the mass of the proton by the mass of the electron. Think about it!

5

The Pluperfect Square;
An Ultimate Patio Decor

Be the first in your neighbourhood (and possibly the first in the entire world) to tile your garden patio in a manner not only unique but, until very recently, unknown. What a conversation piece for a slow news day (providing, of course, that you know at least a little about its fascinating story). But what is this story? What is this tiling pattern? And why all the fuss? In order to answer these questions we need to go back at least to the 1920s, and possibly a little earlier.

It was in the summer of 1925 when the question of dividing up a rectangle into squares which are all of different sizes first made its appearance in the mathematics literature. In fact, two specific examples were actually given at that time, and are reproduced here in figure 5, with the lengths of the sides of the component squares expressed in integers as marked on the figures. It is true that the problem of dividing large rectangles into smaller *rectangles* had been mentioned several years earlier, but this is the first known reference to a division into squares—and, after all, squares are rather special, being in a sense the ultimate rectangle. But if this is so then, presumably, the ultimate problem of this kind should involve the dividing of a large *square* into smaller squares, all of which are of different sizes. This last restriction, concerning the fact that all the component squares must be different, is extremely important. Without it the problem becomes trivial—even I can divide a square into four equal quarters; but that is not what it is all about.

The origin of this ultimate problem, sometimes referred to as 'squaring the square', is something of a mystery. The renowned puzzlist Sam Lloyd, in the early years of this century, presented a patchwork quilt problem, the solution of which required the division of a square into smaller squares, but not *all* the squares were unequal. This requirement of all-different sizes is particularly crucial to the problem of 'squaring the square' since, at the time of Sam

Lloyd's quilt puzzle, it was not even known whether such a separation into all-different squares was possible, no matter how many smaller squares were involved. In fact, one of the very first references in print to the problem, which appeared in 1930, was a proposition (communicated by Professor N N Lusin, of Moscow University) that it is not possible to decompose a square into any finite number of smaller squares with no two equal.

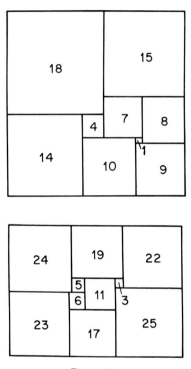

Figure 5

The first emphasis in attacking the problem of 'squaring the square' was therefore to try to establish a general proof that a solution did or did not exist. In fact, such a proof was never achieved, the first positive step being taken by a group of four mathematicians at the University of Cambridge with the discovery of an actual example. This first 'perfect' square, as solutions of the squared-square problem came increasingly to be known, was discovered in 1938, and was decomposed into 69 smaller squares. It was therefore referred to as a 'perfect square of order 69'. This example was followed the next year by a second (discovered by a German group) and this one, which was in fact the first to be published, was smaller than the Cambridge one in the sense that it was decomposed into fewer smaller squares; 55 to be exact.

At this time there was no well-defined method of producing perfect squares, so that the approach was one of a great deal of effort and a 'little bit of luck'. However, more methodical progress was being made on the related, but simpler, problem of 'squaring the rectangle'; that is, dividing up rectangles into squares with no two equal. In that context a general method was obtained for constructing all the squared rectangles of any given order, where the 'order' is the number of squares into which the rectangle is divided. Although we shall not be able to understand the method in any detail, it consisted of relating the problem to an equivalent one involving an electrical network of a particular type called a c-net. It was shown that every squared rectangle could be derived from a c-net in a prescribed, if somewhat tedious, manner. In this way it was quickly established that the smallest possible squared rectangles were of order nine (that is, contained nine squares) and that there were just two of them. Unfortunately neither of these rectangles had equal sides (i.e., were squares) so that the smallest perfect square, which we shall refer to as the pluperfect square, must be at least of order 10.

It therefore turns out that one of the two squared rectangles which were already known in 1925 (the upper one in figure 5) is a squared rectangle of lowest possible order. And although it is not itself a square (but note that, with sides of 33 and 32, it is enticingly close) a general, if rather inefficient, procedure for searching for the pluperfect square is now established. It is to locate all the squared rectangles of each order and to check them out to see if any just happens to be a square. This sounds just fine and dandy, but there is a problem, and a big one at that. You see, the time and effort required to find *all* the squared rectangles of a particular order increases extremely rapidly with the numerical value of the order; 9, 10, 11, 12, ... and so on. In fact, in the pre-computer era of the 1940s, only the complete sets of squared rectangles up to order 13 were known and (alas!) none of them happened to have equal sides. The pluperfect square therefore remained unlocated with an order somewhere between 14 and 55.

With the development of the electronic computer in the 1950s it was only to be expected that the procedure for examining c-nets (and thereby for deducing sets of squared rectangles) should be adapted for machine use. By the year 1960 a Dutch group had succeeded in using computer speed to list all the squared rectangles up to order 15 (that is, composed of up to 15 squares) but still, none was a perfect square. To give you an idea of the magnitude of the task at hand, there are no less than 2609 simple squared rectangles of order 15. The word 'simple', in this context, implies that the rectangle does not contain within it any smaller rectangle which is also made up of squares. If such a smaller rectangle does exist, then the large squared rectangle is said to be 'compound'. Obviously, by definition, the smallest squared rectangle of figure 5 must be simple. However, if you think about it a little, it does not necessarily follow that the smallest squared square (that is, the pluperfect square) is simple, although it might be.

The emphasis on 'simple' squared rectangles results from the fact that from every simple squared rectangle it is trivial to derive a whole infinite family of related compound rectangles, which are consequently of no great intellectual challenge. For example, from any given squared rectangle, simple or compound, a larger compound one can be formed by merely adding to it a square with side length equal to one or the other sides of the original rectangle. Moreover, it is apparent that this technique of generating new compound rectangles can be continued indefinitely. Equally apparent, however, is the fact that none of these compound rectangles can be squares (since they involve adding squares to an existing rectangle). Does this mean that we can forget about compound rectangles in our search for the pluperfect square? Unfortunately not! You see, not *all* squared rectangles are of the above type, and compound perfect squares are certainly possible. This adds another complication to our quest. Since the general c-net method generates only all the simple squared rectangles, a 'low-order' compound pluperfect square could elude our c-net. Although this was not considered likely (i.e., there was a general feeling, though no proof, that the pluperfect square would be simple) it was obviously something to worry about at the appropriate time.

Using the c-net method, and ever increasing computer speed, all the *simple* squared rectangles up to order 18 had been generated by the early 1960s. A complete list of the number of simple squared rectangles is shown below:

Order	Number	Order	Number
9	2	14	744
10	6	15	2609
11	22	16	9016
12	67	17	31427
13	213	18	110384

well over 150 000 squared rectangles and not a single one with equal sides! In 1962 the list was extended to order 19; but still there was not a single perfect square among them. By this time, therefore, it was clear that the pluperfect square would (if it was simple, as expected) contain at least 20 smaller squares within it.

In actuality, rather more than that was now known, since less methodic activity by enthusiastic amateurs picked up after the Second World War, encouraged by the order-69 and order-55 successes of the immediate pre-war years. Although this did not take place in any systematic fashion and was, perhaps, as much of an art as a science, it did meet with further success which gradually (through the discovery of actual perfect squares) reduced the maximum possible order which the pluperfect square could have. In this context, the most noteworthy discovery, by far, was made in 1948 by T H Willcocks, an employee of the Bank of England in Bristol, UK. Being a chess enthusiast and an amateur mathematician, he maintained an active interest in recreational mathematics and, in this manner, became aware of the problem of 'squaring the square' and of the earlier achievements in this field. However, he made no use of the growing catalogue of squared rectangles, but constructed his own stock. From them, by using various ingenious techniques of his own design, he eventually managed to discover several perfect squares, the smallest of which was a compound perfect square of order 24. The magnitude of this achievement can only now, with full hindsight, be fully appreciated. Willcocks' compound perfect square of order 24 is now known to be the very-lowest-order compound perfect square which exists. It was also, until the moment of discovery of the pluperfect square itself in 1978, the lowest-order perfect square known. It is therefore quite worthy of reproduction here and is shown in figure 6.

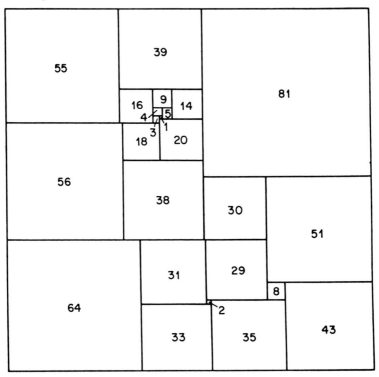

Figure 6 The smallest compound perfect square.

But let us now return to the year 1964 and rejoin the search for the ultimate lowest-order perfect square. It had to contain at least 20 smaller squares and, by virtue of Willcocks' amazing discovery, could contain at most only 24 such squares. But where was it, and what did it look like? The general procedure of squaring rectangles had ground to a halt since the computers available in 1964 were just not capable of generating all the solutions for order 20 in an acceptable time span or at a tolerable expense. The next significant step in the search for perfect squares came from the introduction of a completely new method.

This new method involved the preparation of squares which were divided into unequal smaller squares except for one remaining rectangle. These objects were called 'deficient squared squares' and were easy to construct by a general method which also used an electrical network analogy. After preparing these deficient squared squares, a search was then made in the rapidly expanding catalogue of known squared rectangles to see if any existed with the ratio of side lengths necessary to fill the 'hole' in the deficient square. If so, a new perfect square had been formed. We note that all these perfect squares were, by the method of their formation, of the compound rather than simple category. The new method therefore nicely complemented the earlier one which located only simple perfect squares.

The method, which was first introduced in 1963, quickly led to the discovery of more than 20 new compound perfect squares of low order (less than order 29) the smallest being two of order 25. The procedure was programmed for computer, and many hours of computer time were used in an effort to beat the square shown in figure 6. But no example smaller than this order-24 square (which Willcocks found with just paper and pencil) was ever found, in spite of all the effort. However, by use of this method it is possible to create compound perfect squares of medium or high order with relative ease (even by hand). As a result, little significance is now attached to higher-order perfect squares in spite of the fact that their very existence was in question until only a few decades ago. They can be generated in their tens of thousands; more than 2000 are now known with order 33 or less. The smallest still remains the Willcocks square, next come two of order 25 and then 13 of order 26.

In the late 1960s the focus of attention returned once more to simple perfect squares. These, you will remember, are those which do not contain within them any smaller squared rectangle. By this time several larger simple perfect squares (with order down to 31) had been discovered, but not via any method which could be expanded for more general use. In 1967 the PhD thesis of John C Wilson of the University of Waterloo, Ontario, Canada, set out the first general method which could be used to generate simple squared squares. It did not claim to be able to generate *all* possible simple squared squares of any particular order, but it quickly increased the numbers of such squares known by leaps and bounds. In fact Wilson's techniques, which were graphical in nature, soon succeeded in lowering the order of the smallest known simple perfect square down to 25. To be more specific, Wilson found no less than five simple perfect squares of order 25 and as many as 24 different examples of order 26. But still Willcocks' order-24 square stubbornly refused to be beaten. Could it possibly be the pluperfect square after all?

By the year 1978 the situation had changed little and efforts seemed to be stalled. The lowest-order perfect square known was still T H Willcocks' compound square. Some 10 perfect squares of order 25 were known (eight of them simple and two compound) and 41 perfect squares of order 26 (28 simple and 13 compound). Of the 13 compound squares of order 26, one even contained *two* separate smaller squared rectangles and is thought to be the smallest perfect square of this kind. Computer efficiency had improved further over the years and the complete set of perfect rectangles of order 20 had now been completed, but yet again without the appearance of an equal-sided one. All this therefore still left open the question of whether perfect squares of order 21 to 23 existed.

The answer finally flashed across the computer screen like a bolt of lightning on the night of March 22 1978. It was at the Twente University of Technology in Enschede, The Netherlands. There Dr A J W Duijvestijn was employing a new and highly sophisticated computer program in an effort to push the 'brute force' method of constructing all possible simple perfect squared rectangles and testing them for equality of sides to order 21. Thousands of perfect order-21 rectangles had already been found, and the checking for equality of sides was progressing in an orderly fashion when suddenly there it was!—a beautiful 21 element perfect square. We show it in figure 7 where, as you can see, with a side length of 112 units, it has a largest component square of 50 by 50 and a smallest of 2 by 2. After 40 years of effort the smallest simple perfect square had been found.

But questions remained. Firstly, was it unique? That is to say, were there any more order-21 perfect squares among the thousands of perfect rectangles which the computer continued to unearth? Following many more hours of careful analysis this question was settled. After completing the derivation of all order-21 squared rectangles, only this single example of a simple perfect square existed. It was indeed unique. But was it also the pluperfect square? Duijvestijn was using a technique which was capable only of locating simple perfect squares and, although almost everyone with an opinion seemed to expect the pluperfect square to be simple, a nagging doubt remained. Could a smaller compound perfect square possibly exist? Fortunately, the doubt did not last long. Later that same year a proof was presented that the smallest compound square must have at least 22 components. Duijvestijn's square of figure 7 is indeed the pluperfect square. It holds a truly unique spot in the geometrical world. No other square will ever be found which is made up of as few, or fewer, different smaller squares. Finally, in 1982, it was established that there are no compound perfect squares below order 24, and that there is one, and only one, compound perfect square with 24 components, namely Willcocks' perfect square of figure 6.

The task of categorizing all of the lower-order perfect squares is still far from complete at the time of writing. In particular, the construction of complete sets of simple perfect squares has still not been accomplished by any efficient method which might speed up the process. The production of compound perfect squares has progressed more methodically but, even here, the computer time needed to locate *all* possible examples of a given order is still too long to progress beyond the few smallest orders 24, 25, 26, ...

From our point of view, however, the story is complete. Both the smallest simple perfect square (figure 7) of order 21 and the smallest compound perfect square (figure 6) of order 24 are now known. Each is unique in the sense of

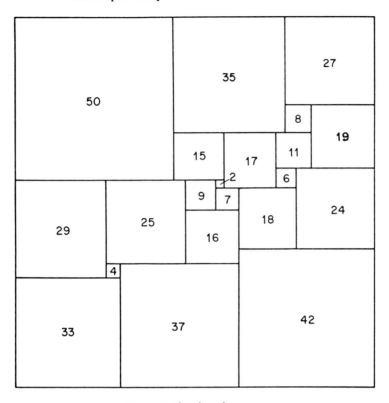

Figure 7 The pluperfect square.

being the only example of its order. Any stencil prepared in either of these configurations could easily be used to imprint a concrete patio floor. Surely the result of more than 40 years of concerted effort by some of the world's most competent mathematicians (both amateur and professional) is worthy of some commemorative effort on your part. Let the patio begin!

Before we take our leave of the pluperfect square it is interesting to note that the existence of perfect squared squares has also led to a very simple solution of a related problem; namely, is it possible to tile an infinite plane with squares, no two of which are the same size? The problem is very nearly solved by setting up a whirling Fibonacci spiral with squares of side length 1, 1, 2, 3, 5, 8, 13, ... as shown in figure 8. If the spiral is continued forever it will obviously eventually cover the infinite plane with squares whose side lengths correspond to the infinite Fibonacci series discussed in an earlier chapter. There is just one problem. The Fibonacci sequence starts off with two ones. This means that the pattern contains just two squares (the smallest ones in the middle of the spiral; labelled with a 1 in figure 8) which are of the same size, and the problem asks for all-different-sized squares. More than 40 years ago, when no squared squares were known to exist, we would have been stumped at this point; so near and yet so far. But now we can partition one of these two equal squares into any of the thousands of perfect squares which are now known, and the tiling problem is solved.

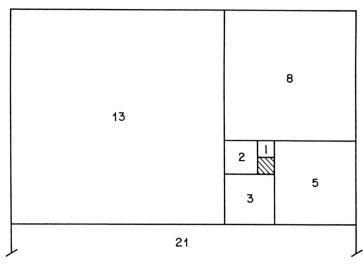

Figure 8

May I now leave you with an unnerving suggestion? Having successfully squared the square, how about 'cubing the cube'? By this I mean dividing up a cube into smaller cubes, no two of which are the same size. What, in particular, does the pluperfect cube look like? Fortunately for you the answer is already known. It is not possible to cut up a cube into *any* finite number of smaller cubes of all different sizes. The proof is one of the most beautiful in all of geometry. If this situation disappoints you do not despair. The three-dimensional version of the tiling problem is still, to my knowledge, unsolved. Spelled out in full this problem is 'can all three-dimensional space be filled with cubes no two of which are the same size?' Please feel free to ponder the situation.

6

The Trouble with Euclid's Fifth

Euclid, let us face it, would not be included by many in a list of most popular authors. We still associate him far too closely with the trials and tribulations of high-school geometry and, after such an experience, are not inclined towards forgiveness. Nevertheless, Euclid's most famous work, a treatise called 'Elements', was used as a text for some two thousand years after it first appeared. Even today, a modified version of its first few chapters forms the basis of all school geometry texts. Both he and it are therefore perhaps worthy of our attention for a few brief moments.

We are all vaguely aware of the fact that Euclid was the 'father' of geometry in the sense that it was he who first set out the explicit 'rules of the game'. Less well known is the enormous controversy which persisted over the centuries concerning these rules. Even less known is the fact that it was just these controversies which finally led to the development of completely new geometries; geometries which were invaluable in assisting the birth of 20th century physics and, in particular, of that science of time and space known as general relativity. But all of this seems a far cry from the work of the ancient geometers who first set out to formulate a science of the properties of space and measurement by carefully defining five 'axioms', which mean self-evident truths.

Now these self-evident truths were meant to be just that; so obviously true that they were beyond question. They could not be proved in any formal sense since, until they were stated as the 'rules of geometry', there was no framework within which to prove anything. Nevertheless, to the early Greek geometers they were far more than a set of non-contradictory rules; they actually represented the real properties of real space to the extent that the latter were self-evident. Let us now take a look at these rules to see if we can spot any 'truth' which might be a little less self-evident than the rest.

The first axiom was that there is exactly one straight line which connects any two distinct points. Our first question here might be 'what do you mean by a straight line?' I suppose that we all have a pretty good idea what a straight line is but, for our game of geometry to be precise, we must be careful to leave no room for misunderstanding. We therefore define this straight line to be the shortest distance between the two points in question. Since the relative lengths of lines can always be compared (at least in principle), this provides a practical method of establishing 'straightness'. Fair enough! On to the next axiom.

The second axiom states that 'every straight line can be continued endlessly'. There is a bit of a problem here, since we can never actually go out towards infinity to see whether this statement is necessarily true for the space in which we actually live. Nevertheless, if you try to picture in your mind a universe of empty space, it is difficult to avoid the implication of 'going on endlessly'. How can you possibly envisage a boundary of empty space? Euclid's assumption does seem the more comfortable concept to live with.

The third axiom is that it is possible to draw a circle with any centre and with any radius. At first sight, this axiom has an odd look about it. What statement is it trying to make about the properties of space? Well, the implication is that the *local* properties of space (such as the ratio of the circumference of a circle to its radius) do not change as you move from one part of space to another. Once again this seems to be in accord with experience. Circles in China (and presumably on the moon) have the same properties as those in the United States.

The fourth axiom is that 'all right angles are equal to one another'. This statement does not make any sense until we have spelled out precisely what a 'right angle' is in a purely geometric sense; so let us first do this. When two straight lines meet, they define four angles (two equal pairs) at the crossing point. If the lines are oriented so that the four angles are all equal, then each angle is said to be a 'right angle'. The axiom then says that these equal crossing angles will have the same size no matter which particular pair of straight lines we are concerned with, or where in space this crossing takes place. Yet again it is difficult to find fault with this assumption.

Finally, therefore, we come to the fifth axiom, which is the centrepiece of our present story. The original statement of this by Euclid himself seemed rather cumbersome but, for the record, we shall first give it in the original form. It proceeds as follows: if two straight lines lying in a plane are met by another straight line, and if the sum of the inside angles on one side is less than two right angles, then the lines will meet if extended sufficiently on the side on which the sum of the angles is less than two right angles. If you draw a picture, the statement, in spite of its length, becomes quite clear. It implies, in particular, that if two lines (and lines always imply straight lines in the present context) are crossed by a third, which makes exactly a set of right angles with each, then the two original lines will never meet.

Now, straight lines which never meet are called 'parallel' by definition. It follows that the fifth axiom (or 'Euclid's fifth' as it is often called, possibly in friendly allusion to the Amendments of the United States Constitution) can be restated using the notion of 'parallel'. In this form, usually attributed to the Scottish mathematician and physicist John Playfair (1748–1819), it is sometimes referred to as Euclid's Parallel Postulate and takes the much simpler form 'only one line can be drawn through a given point parallel to a given line'. From it can be proven all sorts of familiar recollections from school geometry such as 'the angles of a triangle add up to 180 degrees' (or two right angles as Euclid would have more formally stated it). But is it true for the space in which we actually live? Euclid's fifth axiom, like the second one, makes a statement about infinity. If parallel lines *never* meet it implies that we know something about the behaviour of space infinitely far away. However, unlike the second axiom which required only that space continue forever, the fifth implies that we actually know a very precise property of space outside any region which we can possibly check out experimentally.

The modern approach to this dilemma would be simply to say 'don't worry, just let the above axioms *define* the space in which we are going to play our game of geometry'. Since Euclid first came up with the rules, we call it Euclidean space in his honour. However, we now have a considerable benefit of hindsight concerning other possible 'spaces', as we shall set out below. The early Greeks did not. They were concerned with measuring 'real space', that is the space in which we and the stars and the galaxies dwell. The all-important question was 'does the fifth axiom hold for the empty space in which all the matter in the universe is contained?' Spelled out in this fashion the question was not at all easy to answer in a convincing manner. Certainly, as far as a small local region of space is concerned, parallel lines do not seem to have any tendency to come together or to move apart. But how confident can we be that something of that kind might not slowly begin to happen if we could follow the lines far enough?

There certainly seems to be a practical reason for doubt. Nevertheless, for many centuries after Euclid, the general consensus was that the fifth axiom just had to be true. There were two sorts of arguments presented. The first was a religious one. The empty space into which all of God's creations had been placed had to be perfect in His sight. As such it would hardly be likely to contain a vilely converging and diverging morass of straight lines. Euclid's fifth axiom defines by far the simplest and most orderly of spaces which can exist. The divine beauty of creation therefore requires that this be the 'true space'; God would not have botched his work. The second argument, which gained much support in the 18th century, was that the mere idea of an infinite space was necessarily a creation of the mind. Since the mind was not able to imagine a non-Euclidean space, the argument continued, it followed that the fifth axiom must hold. This is a very interesting point of view, but is now unfortunately not one which it is possible to adhere to since, as we shall soon see, it is no longer at all difficult to picture in one's mind a non-Euclidean universe.

But let us not get too far ahead in the saga of Euclid's fifth. There was, among many mathematicians and geometers of the early years, a much less philosophical point of view. It was a feeling that the fifth axiom was not an independent statement at all, but that it somehow must be provable from the first four. If this were true, then it would end once and for all any doubts about the geometry of space in the far reaches of the universe. Scores of erroneous 'proofs' appeared over the years, always to be demolished by closer examination. Even some of the world's greatest mathematicians tricked themselves into believing that they had finally proved Euclid's fifth. For example, a story is told of the famous 18th century French geometer Joseph Louis Lagrange who, at one point in his career, was convinced that he had proved that the angles of a triangle add up to two right angles, just by using the first four axioms. It is said that in the middle of a lecture to the French Academy on his 'proof' he suddenly stopped, muttered 'I shall have to think this over again', and abruptly left the hall.

Finally, in spite of all the claims to the contrary, it was established, beyond any shadow of doubt, that Euclid's fifth could not be proven from the other axioms. It was a completely independent statement. Although the year was 1868, 'proofs' of the parallel postulate continued to appear well into the 20th century. In fact, so great was the obsession that Euclid's fifth just had to be true that, over the years, many perfectly good non-Euclidean systems of geometry were developed time and time again only to be discarded out of an obstinate worship of the Euclidean ideal. On the other hand, it is easy to sympathize with the 'Euclideans' since some of the consequences of dropping Euclid's fifth from the rules of geometry lead to findings which are extremely 'difficult to swallow'. For example, if you draw two perpendiculars of equal length from the same side of a straight line, and then join the ends to complete a rectangle then, without Euclid's fifth, one can only establish the presence of two right angles. The other two angles can only be proven equal; they can just as happily be less than or greater than a right angle as be equal to a right angle. But that is absurd, I can almost hear you saying. How can a construction like that possibly give anything but right angles? We must use our common sense.

Just how far that 'common sense' can lead us from the truth is most easily demonstrated by a visit to Flatland. Flatland is a universe which contains only two dimensions instead of the three which are familiar to us. The Flatlanders, who inhabit Flatland, are also naturally two-dimensional folks, and are as ignorant of the existence of a third dimension as we are of a fourth. Their world is measured solely by east–west and north–south coordinates.

In Flatland life proceeds quite happily, and Flatland geometers have constructed their own rules of geometry in order to measure the properties of the space in which they live. In fact, the five axioms of Euclid suffice just as well for Flatland geometry as they do for us, so that Flatlanders are also confronted with the problem of Euclid's fifth. But now there is a big difference. We three-dimensional people possess a god-like advantage over

Flatlanders in that we can easily imagine the existence of a third dimension. To the poor Flatlanders, who have never even considered the possibility of such a thing, the third dimension is utterly incomprehensible. We, on the other hand, can actually see what the three-dimensional shape of their universe is.

Let us suppose, for example, that the Flatland universe is actually the surface of a large sphere, like a perfectly smoothed earth's surface. Doing geometry on a local scale is no problem for them. Everything seems quite Euclidean. In particular, drawing two equal-length perpendiculars from the same side of a straight line and joining the ends appears to give just as good a 'four right-angle' rectangle for them as it does for us. But now we have (to Flatlanders) that god-like ability to imagine a dimension beyond their experience. We know that a straight line in *their* space (that is, the shortest distance between two points on a sphere) is actually a small arc on what we see as a 'great circle' of the sphere which is their universe.

Being a god in the universe of the Flatlander, you (the reader) can now actually carry out their 'rectangle' drawing experiment on a scale which to them is unthinkable. Find yourself a ball. Its surface is the Flatland universe. It looks quite small to you; but then you are now a god. To the Flatlander this ball is possibly billions of Flatland light-years across. Just take a pen or pencil and draw a 'straight' line on the ball (that means straight in Flatland—the shortest distance between two points on the ball moving along the surface) and raise two perpendiculars of equal length on the same side of this line. Now join the two ends to complete a Flatland rectangle, and what do you observe? The angles formed are equal, but they are not right angles. They are, in fact, larger than right angles. It seems clear that Euclid's fifth is not valid in Flatland. Indeed, it is easy to see that *all* straight lines in Flatland eventually meet. There are no parallel lines at all. Lines which look parallel on a local scale are like lines of longitude on earth; they eventually meet at the poles. But what about lines of latitude, you may say, they never meet. True, but they are not straight lines in Flatland. Except for the equator, there is always a shorter distance between two points on a line of latitude than the latitude line itself.

We three-dimensional beings do not find this non-Euclidean behaviour of Flatland to be at all puzzling. We would say to the Flatlanders 'your universe is not really flat; it is the surface of a sphere and only appears flat to you because the diameter of your spherical universe is so immense that its curvature in your locality is far too small for you to measure'. They would likely respond 'Sphere? Whatever in Flatland is a sphere?' And we would smile knowingly to ourselves—the concept of a third dimension is simply beyond their imaginative powers.

But now let us raise the discussion by one dimension and sense our own intellectual limitations. Who is to say that, if we were able to construct geometrical figures over unimaginable distances in our own universe, something akin to the Flatlander experiment might not occur? Might we not obtain rectangles with angles which add up to more than four right angles? And what if we did? Would it be because our universe was really curved in the fourth dimension? Could we perhaps be living in a space which is really the 'surface' of a four-dimensional sphere? But what on earth is a four-dimensional sphere, you ask? Do we not now sound a little like the Flatlanders, experiencing the problem this time from the side of the less intellectually capable species?

Looking once again at Flatland, we can now also get some idea of what is required for Euclid's geometry to be valid. The Flatland universe would be Euclidean if, from the point of view of three-dimensional beings, it was truly flat (that is, a plane). If the Flatlanders merely adopted all of Euclid's axioms in order to *define* their space, they would be supposing that their universe contained no curvature at all in the third dimension; not that they would truly appreciate what that statement meant. In the same way, if we merely adopt all of Euclid's axioms in our three-dimensional world, then we are assuming that our space has no curvature whatsoever in the fourth dimension. We, in turn, find that statement a bit tricky to understand in any physical sense.

Now clearly if our space *is* curved in the fourth dimension in any manner at all, then Euclid's fifth will not be valid. Since we cannot in practice follow supposedly parallel lines all the way out to infinity, any tests for 'Euclidity' must necessarily take some other form. What might these be? Well, it is easy to show that, if our space is very slightly curved in the region where we live, then it should (in principle at least) be detectable by examining closely the properties of geometrical figures like circles and triangles. You may recall from school days that the angles of a triangle add up to two right angles, and that the circumference of a circle is related to its diameter by that most famous of all irrational numbers pi (or 3.141 592 653 589 793 2 ... where the dots imply a continuation of decimal places to infinity). What the school books possibly did not tell you was that both these statements are true only if we adhere to the truth of Euclid's infamous fifth. This therefore suggests that we look very closely at the triangles and circles around us.

In slightly non-Euclidean space the angles of a triangle will add up to a little more or a little less than two right angles, and the ratio of the circumference to the diameter of a circle will be a little more or a little less than pi. Take out your ball once more, that universe of the Flatlanders, and draw a Flatland triangle or a Flatland circle and test them for yourself. In Flatland we find that the angles of a triangle add up to more than two right angles and that the circumference of a circle with a unit-length diameter is less than pi. The deviations become very large if we can construct triangles and circles which are almost as large as the universe itself, but become very small on a local scale. We might also note that the mighty theorem of Pythagoras for a right-angled triangle with side lengths a, b, and c (namely $a^2 + b^2 = c^2$) also fails. Since none of these strange events has ever been observed to occur in our local vicinity of three-dimensional space (to the accuracy with which it is possible at present to measure them) we must presume that any curvature of our space in the fourth dimension must be very small indeed, at least on a local scale. But then we are also very small compared with the size of the galaxy or, even more ambitiously, with the size of the visible universe. Who is to say what really happens to empty space far enough 'out there'?

One particularly interesting consequence of the idea of having a curved or non-Euclidean universe is that it makes it much easier to imagine a boundless space without having to confront the concept of infinity. To grasp this idea it is best to go once more to Flatland, since there we are as gods. The idea of an infinite universe is just as baffling to the Flatlander as it is to us. He thinks that he lives in one, even though to us it is just the surface of a ball and, in the three-dimensional experience, there is nothing perplexing or infinite about that.

Although we know that Flatland is really a three-dimensional spherical surface, there is no way in which we can physically communicate this idea of a third dimension to the citizens of Flatland. As far as they are concerned their land seems to be locally Euclidean (i.e., a plane surface). They possess no instruments accurate enough to detect third-dimensional curvature in their own back yard. To them their universe is a plane and it is essentially impossible for them to imagine how it can be anything but infinite in size. 'How could it possibly be otherwise?' they ask, 'What kind of barrier could conceivably mark the end of the universe, and what would be beyond it?'

But suppose that one particularly robust and adventurous Flatlander should decide to set out on a trek to test this theory; just in case there really is an 'end of the universe' out there. What would happen? Well, from a three-dimensional point of view the answer is obvious. Starting out along a straight line in Flatland our traveller will imagine himself moving forever away from his starting point towards the edge of the universe. In three dimensions, however, we see him, like a Flatland Ferdinand Magellan, gradually circum-navigating his spherical universe. Should he persist long enough then one day, to his complete astonishment, but in accord with *our* every expectation, he will find himself back at his starting point. 'This is crazy', he will think. 'I have travelled all the time in a straight line, moving ever farther from home. Yet, just when I feel sure that the universe does indeed go on forever, so that I may as well turn around and wend my weary way back home with the news, I find that I *am* home. I must have messed up the navigation somehow and, like a person lost in a fog, travelled in a circle'. But no, we assure him, his line of travel was a true straight line in Flatland, deviating neither to the left nor to the right. The explanation is that the Flatland universe is curved in the third dimension so that, although it has no boundary or 'edge', it is nevertheless of a quite finite size, this size being the surface area of an (to them) unbelievably gigantic sphere. All this, of course, makes little or no sense to our Flatland traveller, who merely mutters that there is no such thing as a third dimension of space, so 'how could anything possibly be curved in it?'

Let us now, once again, raise our complete picture by one dimension. Suppose, at some time in the future, it becomes possible to set out at or near the speed of light aboard spaceship Enterprise to probe the outer reaches of our own universe. Is it conceivable that a fate similar to that of our Flatland traveller might also befall our Enterprise crew? After travelling directly away from the Earth in a straight line towards the farthest reaches of distant space, might they also suddenly find the vicinity beginning to look a little familiar? Could planet Earth also be 'out there'? If our universe were the three-dimensional 'surface' of a four-dimensional sphere then such an event would be a certainty. 'There is nothing very baffling about that', a four-dimensional observer would say, 'the path which the spaceship Enterprise took was indeed a straight line in three dimensions, but it was curved in the fourth dimension'. 'What do you mean', we reply, 'there is no fourth dimension of space, so how could anything possibly be curved in it?' Oh how easy it is to understand if you are just one dimension larger than the problem!

The idea that our 'real' three-dimensional universe may be curved in some manner in the fourth dimension is not science fiction; it is a real possibility. And the geometry necessary to describe such a situation is just that of Euclid, but with the fifth axiom changed. The first person to take a non-Euclidean geometry seriously was that most famous of all German mathematicians Karl Friedrich Gauss (1777–1855). It is not known for certain when Gauss first created a fully self-consistent geometry without the presence of 'Euclid's fifth', since he never published it, but it is certain that he was in possession of many of the main results well before the Russian Nikolai Lobachevski (1793–1856) or the Hungarian Janos Bolyai (1802–1860) published complete theories in the late 1820s and early 1830s. In these first non-Euclidean geometries an infinite number of lines could be drawn through any point parallel to a given line. Such pronouncements were met with great doubts, misunderstandings and misgivings. The idea seemed to verge on madness. Indeed, Bolyai's father wrote to him and implored him to 'For God's sake give it up (before it) deprives you of your health, peace of mind, and happiness in life'. Surprisingly, the conceptually simpler geometry which holds for a spherical surface in two dimensions and for which no two straight lines can ever be parallel, appeared later (in 1854) and is credited to another German mathematician Georg Bernhard Riemann (1826–1866).

This liberation of geometry from the stranglehold of Euclid's fifth has been described by many as one of the major revolutions in all thought. In particular, in the early 20th century it enabled the genius of Albert Einstein to construct a non-Euclidean physical theory of space and time. Einstein was interested in the motion of material bodies (that is, sticks and stones and the like) and had constructed a theory which applied to objects moving at a constant velocity with respect to others. It was called 'special relativity', and first saw the light of day in the year 1906. In an effort to generalize this theory to include changes in velocity (or accelerations) he realized that gravity, as a physical force, is unique. This is because all bodies which fall freely under gravity travel (in space) along the same path no matter how heavy they are or what they are made of. This implies that gravity is actually a geometrical property of space itself (or more strictly speaking, of space and time together).

Within this new theory (the so-called 'general theory of relativity') gravity causes a departure of space from Euclidean form—that is, produces a 'curvature' of space. From this point of view the planets do not revolve around the sun in response to the gravitational force upon them (as the earlier Euclidean theory of Isaac Newton had supposed) but because gravity actually distorts space itself. The motion of the planets is then seen as a completely free motion in a curved space. And the difference between the Newtonian and Einsteinian descriptions is not simply one of 'two ways of looking at the same thing'. Intriguingly, the two descriptions do not produce *identical* predicted paths for the planets, although the differences are small in the comparatively small gravitational fields experienced in our region of the galaxy. Nevertheless, the tiny corrections of Einstein's non-Euclidean approach to the earlier Euclidean picture are measurable, and were first confirmed by a careful study of the orbit of the planet Mercury, for which the sun's gravitational effects are strongest.

There is, therefore, convincing evidence that the real space in which we live is slightly curved (or non-Euclidean) in our neighbourhood. Most significantly, this curvature also affects rays of light. Since light travels at such a great speed, the effects of gravity upon it are far smaller than upon most material objects. Moreover, such effects are completely absent in the Euclidean space of Newton. It follows that any direct observation of the bending of light by gravity would also confirm the non-Euclidean nature of our own 'three-dimensional' space. And such effects have been seen. In particular, such a bending can be observed by precise instrumentation during an eclipse of the sun, when the images of stars in directions close to the sun's edge appear to shift. An observation of this kind was first made in 1919, a few years after Einstein's prediction of this non-Euclidean event, and was another early triumph of general relativity.

Now although the universe seems to be mostly empty, the amount of cosmic 'matter' (from dust, to planets, stars, and even black holes) which it contains is extremely large, simply because the universe itself is so vast. As long ago as 1915 Einstein realised that the countless numbers of small curvatures of local space caused by gravitational effects may not average out to zero as one progresses through space. If that were the case, then there might eventually be enough overall curvature to 'close' space in a manner similar to that experienced by the two-dimensional Flatlanders in their spherical universe. If so, then the real universe in which we live may, like Flatland, be finite yet unbounded; that is, possess a finite three-dimensional volume just as Flatland had a finite two-dimensional area.

Such a universe could, in principle at least, be circumnavigated by a sufficiently adventurous Enterprise crew. It might even be possible, by looking through a powerful enough telescope, to see the back of one's own head! Do not laugh too heartily; this may sound a bit bizarre, but some astronomers claim that already there is a 'higher-than-chance' proportion of radio stars at diametrically opposite points in the sky. Could it be that these pairs of stars are really one and the same star seen from two opposite directions? If they are, then the suggested 'diameter' of our universe is about 10^{10} light years or, in more earthly units of measure, approximately 60 000 000 000 000 000 000 000 miles.

And if all of this is not bizarre enough for you, it should be remembered that a sphere (or any shape that can be obtained from a sphere by stretching or continuously distorting it in any way) is by no means the only three-dimensional object on the surface of which Flatlanders might dwell. Still more perplexing (at least to the Flatlander) properties can be obtained by imagining their two-dimensional universe to be some other kind of curved surface. Consider, for example, the surface formed by a long strip of paper into which one twist has been inserted before joining the ends. If both surfaces of this strip are thought of as the same 'space' (as if the paper had no thickness or even substance) then a journey once around the loop of this new Flatland changes right-handedness into left-handedness. Make a paper strip and see for yourself.

Now what would a three-dimensional universe analogous to this twisted Flatland be like? In any small region of space it would appear to be just as 'Euclidean' as the space we see around us. But now, any space adventurer who set out to probe the 'edge of the universe' by travelling away from earth in a straight line would not only eventually reach earth again but would find it both familiar yet strangely different. Everything would be the mirror image of what it was when he left. To the traveller, the book which he took with him is still perfectly readable while the rest of the earth's books have become unreadable until held up to a mirror. From the point of view of the non-travelling population it is the traveller's book which has become strangely mirror reversed. The only way out of this dilemma is for our space adventurer to go around the universe one more time and then, as a reward for his perseverance, everything will be back to normal.

Although no-one, to my knowledge, has ever seriously suggested any-thing quite as bizarre as the above 'twisted model' for our own universe, Einstein himself believed that we dwell on the three-dimensional 'surface' of a somewhat roughened four-dimensional sphere; that is, in a universe which has a finite volume but is unbounded. This conclusion resulted from the equations of general relativity on the assumption that the amount of matter in the universe is about the same everywhere in it, always has been, and always will be. However, since Einstein's day strong experimental evidence has accumu-lated that this essentially unchanging nature of the universe on a large scale is not correct. On the contrary it is now thought that the universe began with a 'big bang' and may well end eventually with a 'gib gnab' (or whatever the reverse of a big bang is called). Since the degree of non-Euclidean curvature in various parts of the universe depends on the distribution of mass, the 'shape' of our universe may well be changing with time; as if we did not have enough problems already!

Perhaps only one thing is certain; there is indeed 'trouble with Euclid's fifth'. Local regions of the universe in which we live do not obey the rules of high-school geometry if the measurements are carried out with sufficient accuracy. And what is the moral of all this? It is that intuition is a powerful tool in mathematics and science, but it cannot always be trusted. The structure of the universe, like pure mathematics itself, tends to be much stranger than even the greatest mathematicians and scientists suspect.

7

Clock Numbers;
An Invention of the Master

What is a clock but a counting machine? It counts minutes, it counts hours, and sometimes even seconds. What is more, so long as its source of power (whether old fashioned or modern) remains intact, it goes on counting essentially forever. On the other hand, there is something different about the way in which a clock counts because, even though it counts (ideally) forever, it never seems to get up to any large numbers. For example, every time the number of hours reaches 12 it starts all over again. If it is 2 o'clock and we wait for 12 or 24 hours it tells us that it is 2 o'clock all over again. It seems not to care about how many lots of 12 hours have gone by, even though it has painstakingly recorded them, choosing to 'remember' only the remainder that is left over after 12 o'clock.

All this is so familiar to us that it does not seem at all strange. And yet it certainly leads to some very odd looking arithmetic. Suppose, for example, that it is 8 o'clock and we wish to know what the time will be if we 'add on' 6 hours. The problem is not exactly one which requires superior mathematics and we readily deduce the answer—namely 2 o'clock. Thus, from the clock's point of view

$$8 + 6 = 2.$$

Looked at as a statement of arithmetic, rather than of time, this equation certainly has an unusual appearance. Nevertheless, this 'clock arithmetic' is quite self-consistent and we may quickly verify such other correct statements as $2 - 4 = 10$, $2 + 12 = 2$ and $6 + 38 = 8$. What we are doing is counting in sets of 12 and recording only the remainder.

Quite obviously there is nothing magic about the number 12 in all of this. Most clocks also count both minutes and seconds and again they attach importance only to remainders, although this time they disregard how many sets of 60 have passed by. From the second hand's point of view $8+6 = 14$, just as it does in conventional arithmetic, but $8+62 = 10$.

Although each of these equations makes sense if we spell out exactly what it is we are doing in each case, it is clear that enormous confusion will result unless we devise some simple way to signify just what we intend. For example, when the teacher asks Johnny or Jill to complete the equation $8 + 6 = ?$, which is written on the blackboard, the answer 2 (which makes perfect sense to the hour hand of a clock) is frowned upon by scholastic authority and firmly denounced as 'wrong'. On the other hand, what would normally be thought of as 14 really does become 2 in clock arithmetic, or at least in clock arithmetic according to the hour hand. To make this clear it has become customary to write something like

$$14 = 2 \qquad (\text{mod } 12)$$

where the (mod 12) implies counting in sets of 12 and caring only about the remainder. Mathematicians refer to this relationship in the rather pompous fashion '14 is congruent to 2 modulo 12' and often use a new symbol $\equiv$ instead of $=$ (presumably on the notion that if you make it too simple no-one will be impressed and, even worse, everybody will be able to understand it). We can read it as '14 is the same as 2 on a 12-clock' and understand it more precisely as '14 has a *remainder* of 2 when counting in sets of 12'.

We can quickly get used to the notation by considering an example or two of more general form. Thus $49 = 1$ (mod 8) says that 49 has a remainder of 1 when counting in eights, which is clearly true. Equally obviously $55 = 7$ (mod 8); that is, 55 has a remainder of 7 when counting in eights. Simple enough! However, we can go a little further and introduce the idea of a negative remainder without too great a stretch of the imagination. Clearly, if 55 is seven units more than a complete number of eights (as set out above) it can equally well be thought of as one unit *less* than a completed number of sets of eight or, as a 'clock equation',

$$55 = -1 \qquad (\text{mod } 8).$$

In a like manner we have $99 = -1$ (mod 10).

This is all well and good, you may be saying, but what use is it? Good question! Its use lies in the fact that both sides of a 'clock number' equation can be added to, subtracted from, multiplied or raised to a power, and still remain true. In other words, except for the operation of division (which has to be treated more carefully and will be discussed a little later in the chapter) the two sides of a clock equation are just as equal as if we were dealing with ordinary equations and we may treat them accordingly.

Consider, for example,

$$49 = 1 \qquad (\text{mod } 8)$$

and add 2 to each side. It becomes $51 = 3$ (mod 8), and is quite obviously still true. Just as trivially we could subtract 2 from each side to get $47 = -1$ (mod 8). Perhaps not quite so trivial is the result obtained by multiplying both sides by the same number; say 49. It gives

$$49^2 = 49 \qquad \text{(mod 8)}.$$

Since 49 on an eight-clock is the same as 1, and 49^2 is equal to 2401, the above result translates finally to

$$2401 = 1 \qquad \text{(mod 8)}.$$

The correctness of this finding is easily confirmed by dividing 2401 by 8 and verifying a remainder of 1. But still, you may think, such a statement (though true) is hardly astounding.

It is in taking powers of both sides of a clock equation where the first mind boggling results begin to emerge. Suppose we start once more with $49 = 1$ (mod 8) and raise each side to (say) the 100th power. Since 1 raised to any power, no matter how large, is still 1, we immediately obtain the result

$$49^{100} = 1 \qquad \text{(mod 8)}.$$

Now 49^{100} is a number far larger than the number of atoms in the entire universe. It contains 170 digits when written out in full, more digits than almost any of today's computers can deal with. And yet, from the above clock equation we know immediately that when divided by 8 it has a remainder of 1. We therefore also know, by subtracting 1 from each side, that the equally immense number $49^{100} - 1$ is *exactly* divisible by 8. This you could still possibly verify by straightforward 'number-crunching' (if you had a few months at your disposal and were very fond of doing careful arithmetic) but with clock arithmetic we can just as easily progress to numbers which even the world's fastest computers could never deal with by direct methods.

Starting once more from the same trivial clock statement $49 = 1$ (mod 8), why not raise each side to the 1 000 000th power? No problem! We get

$$49^{1\,000\,000} = 1 \qquad \text{(mod 8)}$$

done in a snap! That number on the left-hand side now contains well over 1 500 000 digits when expressed in decimal form, and would completely fill up several books of this size. Yet we still know that when divided by 8 it has a remainder of 1, and that $49^{1\,000\,000} - 1$ is exactly divisible by 8.

That is all very well, you may say, if a 1 or a 0 happens to be on the right-hand side of the clock equation (since we immediately know the values of 1^n and 0^n for any n-value no matter how large) but is this not a bit restrictive? Well, surprisingly it is not, since one can nearly always arrange to produce a right-hand side of 1 or 0 by using a little bit of ingenuity.

Suppose, for example, that your best friend asked you whether the very large number consisting of a 1 followed by 999 999 zeroes and a 9 was divisible by 13. Your first reaction might range anywhere from 'you must be joking!' to 'with friends like that who needs enemies?'. Nevertheless, with our newly found clock-fashion arithmetic all is not lost. The number in question can be rewritten as $10^{1\,000\,000} + 9$. Since we are asked about divisibility by 13 we evidently want to work with a (mod 13) clock, and the simplest place to start is with an obvious relationship like

$$10 = -3 \qquad (\text{mod } 13).$$

This is merely a statement that, when counting in sets of 13, the number 10 has a remainder of -3. Multiplying this clock equation by itself (that is, squaring both sides) and remembering that -3 times -3 is equal to $+9$, gives us

$$10^2 = 9 \qquad (\text{mod } 13)$$

or (in words) one hundred has a remainder of nine when counting in sets of thirteen. The truth of this statement is easily verified since seven thirteens make 91 and therefore another 9 is needed to reach 100.

Now how can we most simply produce a 1 or a 0 on the right-hand side? Well, how about multiplying the two relationships set out above together? Since 10 times 100 is 1000 (or 10^3) on the left side, and -3 times $+9$ is -27 on the right, this translates to

$$10^3 = -27 \qquad (\text{mod } 13).$$

But any number which has -27 left over when counting in thirteens must also have -1 left over as well since $-27 = -13 - 13 - 1$. It follows that

$$10^3 = -1 \qquad (\text{mod } 13)$$

and this finding still involves numbers small enough to be checked directly. Since 77 times 13 is 1001, 1000 is indeed the same thing as -1 on a '13-clock'.

The important thing is that we have now arranged for a 1 to appear on the right-hand side. Well, it is actually a -1, but that does not matter since we know that -1 raised to any power (say n) is equal to -1 if n is odd and to $+1$ if n is even. Let us now raise both sides of this last clock equation to the power $n = 333\,333$. As it is odd we immediately obtain

$$(10^3)^{333\,333} = -1 \qquad (\text{mod } 13).$$

Since when we raise a power to a power we just multiply the exponents (e.g., $(10^3)^2 = 10^6$ or, in words, a thousand times a thousand is a million) the above can be re-expressed as

$$10^{999\,999} = -1 \qquad (\text{mod } 13).$$

We are now getting close to our target number of $10^{1\,000\,000} + 9$ but we are still not quite there. We could use another power of 10 on the left-hand side. So let us go back to our starting relationship of $10 = -3$ (mod 13) and multiply by it. Once again, remembering that a minus times a minus is a plus, we find that

$$10^{1\,000\,000} = 3 \qquad (\text{mod } 13).$$

It now only remains to add 9 to each side to reach, at last, our final destination, namely

$$10^{1\,000\,000} + 9 = 12 \qquad (\text{mod } 13)$$

which says that the number in question is not exactly divisible by 13 but, when divided by 13, has 12 (or equivalently, since 12 and -1 are the same on a '13-clock' -1) left over. It follows that it is the number $10^{1\,000\,000} + 10$ which is exactly divisible by 13, even though by outward appearances it does not look to be a very likely candidate for this distinction. In the above spirit, armed with clock numbers, it is now possible for you to examine numbers of almost unthinkable size and to test them for divisibility by any number which is not too large for simple manipulation.

We have so far avoided any operations involving division because the rules for dividing clock numbers are a bit more restrictive than the other rules. To start with, one is only allowed to divide both sides of a clock equation by the same number if it does not give rise to fractions. Thus, for example, we *can* divide $12 = 2$ (mod 5) by 2 to get $6 = 1$ (mod 5), which is still obviously true, but a division by 3 to obtain $4 = \frac{2}{3}$ (mod 5) has no meaning, at least within the simplest notational system which we are concerned with in this book. One further restriction also applies. It is that, if the number we are dividing by also exactly divides the grouping (or mod) number as well, then the latter must also be so divided. For example, we can divide the clock equation $15 = 6$ (mod 9) by 3 only in the form $5 = 2$ (mod 3) and not as $5 = 2$ (mod 9). A quick examination of simple relationships like these makes the rule quite clear; after all $5 = 2$ (mod 9) is plainly not true!

Although we have as yet barely touched upon the possible uses of clock arithmetic (and the idea can be applied to much more than numerical calculations) it is already clear that the method opens up a powerful new approach to the study of the properties of numbers. Let us, therefore, set aside a few moments to review briefly the life of their inventor, Karl Friedrich Gauss. Gauss was a child prodigy. Born in Brunswick, in what is now West Germany, in the year 1777, it is said that he first demonstrated his unique mathematical genius at the tender age of eight when his teacher, in order to keep the students occupied for a while, asked them to add up all the numbers from 1 to 100. There is, of course, a formula for problems like this which the teacher knew but the children did not. He therefore expected to get an hour-or-so's peace and quiet out of this exercise while the students carefully performed their arduous task. Possibly one or two would actually complete the chore without error, although it hardly seemed likely. To his great surprise, young Karl Gauss immediately walked up to the front of the room and presented the correct answer: 5050.

Gauss, it turned out, knew the formula too. But unlike the teacher, he had not learned it; he had quickly deduced it for himself. The trick is not difficult once you know it: you simply add 1 to 99 (to get 100), then 2 to 98 (100 again), then 3 to 97, 4 to 96 and so on all the way to 49 to 51, getting 100 at each step. That makes 49 lots of 100 which, when added to the 50 in the middle and the 100 at the end delivers the correct answer of 5050. Luckily for the subsequent development of mathematics the teacher recognized this event as a sign of genius in the boy, and thus began the career of the man considered by many to be the greatest mathematician of all time.

By the age of 18 he had already established the impossibility of constructing the regular heptagon (that is a seven-sided figure with all sides of equal length and all interior angles equal) with a ruler and pair of compasses alone. This was something which mathematicians and geometers had been attempting in vain for more than 2000 years. For his doctoral thesis he submitted a proof concerning the number of solutions which algebraic equations could have. This theorem is still called 'the fundamental theorem of algebra' and had also eluded the best mathematical minds for centuries. In his spare moments he turned his mind to astronomy and, in fact, he was the director of the observatory at the University of Gottingen as well as the professor of mathematics at that same institution from 1807 until his death in 1855. Nevertheless his principal work was in mathematics and theoretical physics. In recognition of his work in the latter field, the unit of magnetic field intensity is today called the 'gauss' and perhaps the most fundamental theorem of electrostatics is still known as 'Gauss' Theorem'. His work also embraced the field of statistics in which today the most basic and best known of all probability distributions is known as (what else) a 'Gaussian'.

Gauss' most important work on the theory of numbers was the book *Disquisitions Arithmeticae* which appeared, when he was still but 24 years old, in the year 1801. It was in the opening sections of this book that Gauss first introduced the theory of congruences, those clock numbers that ever since have put their stamp on virtually all research in number theory. In the following sections many problems, some of them previously attacked without success by earlier generations of prominent mathematicians, here received their solution for the first time. The extent of his genius may be judged from the fact that in some of these cases he presented as many as three quite different proofs of the same theorem. In other words, what no-one else had been able to prove at all, Gauss proved once and then twice more for good measure.

After that short 'aside' concerning their inventor, let us now take a look at the way clock numbers can be used to establish proofs in number theory proper. Do not be alarmed by this rather formal sounding context; clock equations (or congruences, as they are more usually known) are just as easy to understand here as they were before. With their assistance we can, for example, demonstrate one of the most famous of all number theorems, called Fermat's Little Theorem (after the French mathematician Pierre de Fermat, 1601–1665, who first established and proved it, although his original proof was never published). This theorem is fun since it leads to a method whereby a number can be proven to have factors *even though none of them is known*.

Consider an arbitrary prime number, say 7. Write down all the integers

1, 2, 3, 4, 5, 6 smaller than it, and multiply each by 2. We obtain the new sequence 2, 4, 6, 8, 10, 12 which, if we count on a 'seven-clock' (or, more formally, modulo 7) translates to 2, 4, 6, 1, 3, 5. This is merely the original set in a different order. It follows that, when counting 'modulo 7', the numbers 1 to 6 multiplied together (which is usually written in the shorthand fashion 6! and called 'factorial six') must be exactly equal to the numbers 2, 4, 6, 8, 10, 12 multiplied together. But each member of this second set is just a factor of two times its corresponding member in the first set. It follows, since there are six members in the set, that the numbers 2, 4, 6, 8, 10, 12 multiplied together must also be equal to 2^6 times 6!. We have therefore established that

$$2^6 \times 6! = 6! \qquad \text{(mod 7)}.$$

Now since 6! (by its definition) does not contain any factor which exactly divides the modulo number 7, we can (according to the rules for division set out earlier) divide both sides by 6! to get

$$2^6 = 1 \qquad \text{(mod 7)}.$$

As 2^6 is 64, which is 9 lots of 7 plus a 'remainder' of 1, the correctness of the result is easy to verify by direct calculation and therefore does not yet represent anything particularly worthy of adulation.

The important point is that the method can be generalized. If, for example, we multiply the original sequence of integers 1 through 6 by 3 (instead of 2) the resulting numbers 3, 6, 9, 12, 15, 18 are equal to 3, 6, 2, 5, 1, 4 (mod 7) which, once again, is the original set in 'jumbled' order. The same multiplication argument now generates the congruence

$$3^6 = 1 \qquad \text{(mod 7)}.$$

In exactly the same way we can go on to multiply, in turn, by 4, 5 and 6 to establish

$$4^6 = 1 \qquad \text{(mod 7)}$$
$$5^6 = 1 \qquad \text{(mod 7)}$$
$$6^6 = 1 \qquad \text{(mod 7)}$$

as well. Only when we get up to the 'clock number' itself (in this case seven) does the pattern change since 7^6 (which is just six sevens multiplied together) is exactly divisible by 7 to give $7^6 = 0$ (mod 7).

As long as we count (mod p), where p is a *prime* number, it is not difficult to establish that the set of integers from 1 up to $p-1$ will always transform into themselves (mod p) when multiplied by 2 or 3 or any integer up to $p-1$. On the other hand, if p is not a prime this will not usually happen. For example, for $p = 6$, the numbers 1, 2, 3, 4, 5, when multiplied by 2 become 2, 4, 0, 2, 4, 0 (mod 6) and, when multiplied by 3, become 3, 0, 3, 0, 3, 0. The appearance of zeroes occurs because both 2 and 3 exactly divide into 6. If you experiment a little with a few more primes and non-primes you will quickly become familiar with what is happening.

The final discovery, therefore, is that for any prime number p, and any other number n not equal to p (or an exact number of times p)

$$n^{p-1} = 1 \qquad (\text{mod } p)$$

or, subtracting 1 from each side,

$$n^{p-1} - 1 = 0 \qquad (\text{mod } p).$$

Stated in words this says that the number $n^{p-1} - 1$ is always *exactly* divisible by p if p is prime, unless n itself is also divisible by p. Let us check out a few examples to see how this works. We shall, for simplicity, look first at the numerically easiest examples with n equal to 2. In this manner we quickly verify that $2^2 - 1 = 3$ is indeed divisible by 3 (which is a prime); $2^3 - 1 = 7$ is not divisible by 4 (which is not a prime); $2^4 - 1 = 15$ is divisible by 5 (a prime); $2^5 - 1 = 31$ is not divisible by 6 (not a prime); $2^6 - 1 = 63$ is divisible by 7 (a prime) and so on.

At first sight it looks as though we have here a method of testing for prime numbers. In fact, the Chinese, as long ago as 500 BC expressed the belief that the number $2^{p-1} - 1$ is always divisible by p when p is prime, and never when p is not prime (or is 'composite' to use the proper word). But is this true? We must be careful here. Our proof was established for p equal to a prime. It did not, however, rule out the possibility that $n^{p-1} - 1$ (in general) and $2^{p-1} - 1$ (in particular) might just accidentally also be divisible by p for some occasional non-prime p. For the case of $2^{p-1} - 1$ the Chinese were never able to discover one and, in fact, no such example was found until the year 1819 when it was first noted that the number $2^{340} - 1$ is exactly divisible by 341, and 341 (being 11 times 31) is not a prime.

Numbers which satisfy the congruence shown at the top of this page when p is not a prime are comparative rarities. Nevertheless there are an infinite number of cases and they define the so-called 'pseudoprimes' p. The case with $n = 2$ is the most thoroughly studied. For this case there are 14 884 pseudoprimes smaller than 10^{10} compared with the very much larger number of 455 052 512 real prime numbers less than this same limit. It follows that only about three thousandths of one per cent of the numbers p which pass the

$$2^{p-1} = 1 \qquad (\text{mod } p)$$

test are not real prime numbers. Most of the 'impostors', that is, the $n = 2$ pseudoprimes, are odd. In fact it was not until 1950 that the first even pseudoprime (161 038) was found, and they remain quite difficult to locate.

In addition to the $n = 2$ pseudoprimes, equivalent studies can be carried out for $n = 3, 4, 5, \dots$ and so on. The smallest pseudoprimes for $n = 2$ through $n = 10$ are now known to be 341, 91, 15, 124, 35, 25, 9, 28 and 33 respectively. In fact, the $n = 2$ pseudoprime 341 is the largest 'smallest pseudoprime' for all n up to 100 (although the same value 341 occurs not only for $n = 2$ but also for $n = 15, 60, 63$ and 78 as well). This explains why the Chinese were unable to locate it. You see, $2^{340} - 1$ is quite a large number, containing 103 digits when written out in full decimal form. To calculate this form and then test for divisibility by 341 is a major undertaking. Unless, of course, you know clock arithmetic (which the Chinese did not). But we do and, using the rules as set out earlier in the chapter, the fact that $2^{340} - 1$ is divisible by 341, i.e., that

$$2^{340} = 1 \qquad (\text{mod } 341)$$

can be verified in just a few steps. Give it a try!

Beyond the concept of ordinary pseudoprimes come the 'super-pseudo-primes'. These are numbers p for which $n^{p-1} - 1$ is exactly divisible by p for *all* n which are mutually prime to p. They are called 'Carmichael numbers' and there may be infinitely many of them, although this is not yet known for sure. The smallest is 561, which is $3 \times 11 \times 17$, and it is known that such numbers must be the product (that is the multiplication) of at least three prime numbers. Thus, $2^{560} - 1$ is exactly divisible by 561, as also is $4^{560} - 1$, $5^{560} - 1$, $7^{560} - 1$, and so on over *all* n-values except those divisible by 3, 11 or 17. The next Carmichael numbers in increasing order are 1105, 1729, 2465, 2821, 6601 and 8911, and these are the only ones less than 10 000. There are, in fact, only 1547 Carmichael numbers (or super-pseudoprimes) less than 10^{10} compared with the more than 455 million real primes below this limit.

Although Fermat's little theorem does not (alas!) provide us with a 'watertight' recipe for locating prime numbers, because of the ever-present (though small) possibility of generating a pseudoprime masquerading as a prime, it does enable us to do something almost as impressive. Since any odd number p which does not satisfy Fermat's little theorem cannot possibly be prime, it allows us to establish that certain (possibly very large) odd numbers *must* have factors even though we have not got the faintest idea what they are. There are, in fact, many large numbers of particular interest to mathematicians which, by this and like methods, are known to be composite but for which, as yet, no actual factors have been located.

One final story about Gauss and his clock numbers will suffice to round off this chapter. In our proof of Fermat's little theorem we met the set of numbers 1, 2, 3, ..., $p-1$ where p is prime. Suppose that we multiply all these numbers together to get $(p-1)!$ in the factorial notation. To what would this number be congruent if we counted on a (mod p) clock? Well, let us first try it out for a few of the very smallest primes. For $p = 2$, $(p-1)!$ is 1 and $1 = -1$ (mod 2). For $p = 3$, $(p-1)! = 2$ and $2 = -1$ (mod 3). For $p = 5$, $(p-1)! = 24$ and $24 = -1$ (mod 5); while for $p = 7$, $(p-1)! = 720$ and yet again $720 = -1$ (mod 7). There seems to be a pattern here with $(p-1)!$ wanting to be the same as -1 when counting on a 'p-clock'. You can establish that the rule is indeed quite general by noting that the factors which make up $(p-1)!$ can always be taken in pairs, 1 times $p-1$ giving -1 (mod p) and the others all giving $+1$ (mod p). For example, with $p = 11$ we have 2 times 6, 3 times 4, 5 times 9 and 7 times 8 all equal to $+1$ (mod 11) which, when multiplied by the

−1 (mod 11) of the end-member product (1 times 10) gives the final result. We therefore conclude that

$$(p-1)! = -1 \qquad (\text{mod } p)$$

for all prime numbers p. If, on the other hand, p is not a prime, then two of the numbers multiplied together in $(p-1)!$ must have the product p; that is, must be equal to 0 (mod p). Since zero times anything is still zero, it follows that

$$(p-1)! = 0 \qquad (\text{mod } p)$$

if p is composite. Aha! Here we really do have a watertight test for primeness. The number $(p-1)!$ has a remainder of −1 when divided by p if p is prime, and no remainder at all if p is not prime. The bad news is, unfortunately, that the test is of no practical value since factorials, unlike powers, are not adaptable for easy manipulation using clock arithmetic. That is to say, it takes longer to find the remainder of $(p-1)!$ divided by p, than it does to look for factors of p directly. Nevertheless the general result is interesting since it tells us that there are methods of testing for primeness which do not involve checking out all the possible divisors.

The above result for prime numbers is known as Wilson's theorem and has an unusual history. You see Wilson, or Sir John Wilson, to give him his full title, was a fairly unobtrusive English judge, and few people have ever become immortalized for less reason. He was neither the first to state the theorem nor to prove it. In fact, he never was able to prove it and never published anything about it. Likely it would have amounted to nothing at all if, by chance, he had not mentioned it over lunch one day to a friend of his who happened to be a professor of mathematics at Cambridge University. The professor then published it as a 'speculation', citing Sir John as its originator. It was this publication which caught the eye of the mathematical fraternity and, as a result, Wilson's name became associated with the theorem. Although there is now ample evidence that the result was known to Baron Gottfried von Leibnitz (German philosopher and mathematician 1646–1716) almost 100 years before all of this took place, Wilson's name stuck and is now forever enshrined in all textbooks concerning the theory of numbers.

Even more amusing is the related 'Gaussian' story. Apparently, accompanying the first publication of 'Wilson's theorem' in 1770 was the prediction that this 'conjecture', as it then was, was not likely ever to be proven because there was no known notation for dealing with prime numbers. When this statement was first communicated to Gauss, he proceeded to prove the theorem by 'clock numbers' in five minutes, commenting that what was needed was not notations but notions. In truth, however, Gauss' proof of Wilson's theorem was not the first to be published. That honour went to the French mathematician and astronomer Joseph Louis Lagrange (1736–1813) who presented his proof less than one year after the 'conjecture' first appeared in print.

One final comment on Wilson's theorem is perhaps of some interest to the number addict. The original theorem states that $(p-1)!+1$ is always exactly divisible by p when, and only when, p is a prime number. This same expression is, on very rare occasions, divisible not only by p but by p^2 as well. The only primes less than 10^5 for which this is true are 5, 13, and 563.

8

Cryptography; The Science of Secret Writing

I suppose that as children we have all, at one time or another, made an effort to write down a message in code; a 'secret' message which hopefully can be understood (or deciphered, to use the proper word) only by the intended receiver who has been given the 'key' to the code in advance. Most of us probably used that simplest of all coding systems (or 'encryptions' as the experts call it), the substitution of one letter of the alphabet for another. Suppose, for example, I replace A by B, B by C, C by D, and so on right up to replacing Z by A, then the delightfully uninteresting message 'I HAVE A RED PENCIL BOX' takes on its coded form

JIBWFBSFEQFODJMCPY

and certainly looks mysterious enough at first sight.

Since we could have chosen any letter of the alphabet to substitute for any other, there are a vast number of different schemes of this simple substitutional kind. To be precise, there are 26! (factorial 26) of them, a number whose value is about 4 times 10^{26}. With this almost unthinkably large number of choices for our coding, the chances of anyone discovering the key (or 'breaking' the code) might appear to be extremely remote. And if we wish to transmit only a single message of very short length (such as the red pencil box gem set out above) then this is true so long as we avoid the extremely simplistic patterns such as the A to B, B to C, C to D, etc example of the first paragraph.

Unfortunately, if the same code is required for use in transmitting large amounts of information then, in spite of the size of numbers like 4 times 10^{26}, it is very easy to 'break'. The villain is the non-random nature of the appearance of letters and letter combinations in any particular language. In English, for example, e is the letter which occurs most frequently, q is always followed by u, certain combinations like 'in', 'it', 'the' and 'and' are very common, while others like 'pbv', 'bcd' and 'pxq' do not occur at all. Using these language 'fingerprints', and countless other statistical clues from the pattern in which various combinations of letters arise, it is possible for a 'decoder' to quickly break any code formed by a simple letter substitution scheme. This remains true even if a sequence of meaningless squiggles is used to replace the letters; the pattern of the language still eventually identifies them and our secret messages soon become common knowledge.

A second kind of coding scheme that suggests itself is one which calls for a 'transposing' or shuffling of the letters rather than a substitution. For example, suppose that I write down as my 'key' the sequence of numbers 3, 1, 5, 2, 6, 4. I now transform my message into coded form by moving the first letter to the third place, the second letter to the first place, the third to the fifth place, and so on up to the sixth letter in fourth place (following the number pattern of the key). After transposing the first six letters in this manner, the pattern can be repeated endlessly by transposing the second six letters in the same way, then the third six, and fourth, and so on to the end of the message. Thus, my prototype message 'I HAVE A RED PENCIL BOX' has its first six letters transposed to AIEHAV, its second six letters to DREENP, and so on to the completed form

AIEHAVDREENPLCOIXB.

Although very short messages using this scheme can often not be unambiguously broken (e.g., the coded message, or 'cipher' TEQIU could equally well be QUITE using the key 35412, or QUIET using the key 35421), longer messages are again quite insecure and general methods for their 'solution' were published as long ago as 1878 for ordinary English text, even if the length of the sequence of numbers in the key was completely unknown at the outset. Thus, the simplest transpositional coding scheme, like its substitutional counterpart, provides only an extremely limited degree of security.

What can be done to make the codes a little more difficult to break? Well, one idea is to strengthen the substitutional scheme such that each letter of the original is not always replaced by the same substitute in the coded form. One way to accomplish this is to think of a word (let us choose LOUNGE as an example) and to write it in repeated form below the message to be coded as follows:

I HAVE A RED PENCIL BOX
L OUNG E LOU NGELOU NGE

Since 'I' is the ninth letter of the alphabet and the 'L' below it is the 12th letter of the alphabet, we can code the first letter of the message by adding 9 to 12 to reach the 21st letter of the alphabet 'U'. In similar fashion, for the second letter of the message, H(8) is 'added' to the O(15) below it to produce the coded substitute W(23). If this addition procedure should lead to a number larger than 26, then we simply start the alphabet over again by associating 'A' with 27, 'B' with 28 and so forth. Using this particular 'key' the full message above translates into

UWVJLFDTYJLSOXGPVC

where we have again closed up the spacing between the words to make the decipherment more difficult.

This kind of coding system, often referred to as a Vigenere cipher, obviously does not always replace a particular letter of the original message by the same substitute. Thus, the first 'I' of the message above codes into a 'U' while the second 'I' of the message codes into an 'X'. Surely this would help to confuse anyone attempting to break the code, and make it more secure than either of the first two methods set out before it. Unfortunately, even though this statement may possibly be true, the fact remains that Vigenere ciphers like that set out above can still readily be decoded. In fact, once again a rather general method of attack was developed well over 100 years ago. The weakness is the repetition in the key-word line (for example, LOUNGE-LOUNGELOUNGE... above). For a six letter key like this, common words such as 'and' and 'the' are always replaced by one of only six particular combinations of three letters. Clues like this eventually identify the key and the code is broken. The problem once again is the non-random occurrence of letters and words in the English language (or any other language for that matter).

However, what if we used for the key a text which did not repeat itself—ever? In these 'running-key Vigenere ciphers' the giveaway common words and letters would never repeat in the coded form so that surely, at last, this would lead to an unbreakable code. In fact, for almost three centuries after they were first invented, these running-key ciphers were thought to be just that, completely secure. However, in 1883 a method was described for breaking down codes even of this kind. The clue to their solution again involves the non-random frequency of occurrence of letters and words which now run through both the message and the key. For example, the letter 'E' has a probability of being replaced by another 'E' about 1.69% of the time, compared with a smaller (but still known) probability of being replaced by any other particular letter. Obviously, this kind of code breaking is much more difficult than the others discussed above, and a very considerable length of cipher is required in order for the tell-tale correlations to show up. But sooner or later they do appear and once again the code can be broken. What is more, the development of fast computer techniques since the 1950s has made these running-key ciphers more vulnerable than ever.

What can be done? Well, the weakness of the code is evidently contained in using a piece of English text for the running key. Suppose that the running key was to be made up of a completely random sequence of never repeating letters. In this case, each letter of the text, whether it be the most probable 'E' or the least probable 'Q' or 'Z', is *equally* likely to be replaced in the cipher by *any* other. In such a scheme all inter-symbol correlations or periodicities on which code breaking is based would be totally removed. At last the system would be completely secure. Unfortunately, however, this system is also highly impracticable since it requires one symbol or key (to be exchanged between the sender and receiver in advance of communication) for each and every symbol of text to be transmitted. This, it might be argued, is true for any running-key Vigenere cipher, random or not. The essential difference is that a non-random key of (say) an English text could be taken from convenient locations such as designated parts of well-known (or not so well-known) books. The random key, on the other hand, would have to be carried with, or kept by, the intended receiver and because of its unusual appearance would immediately arouse suspicion if it fell into 'enemy' hands. One possible way around this problem, it was first thought, might be to use a still lengthy, but repeating, random key which could be memorized. However, the periodicity of the key, even though the repeated part is random in its frequency of letter selection, proves to be quite sufficient to enable the code to be broken, given a sufficient length of message.

The inescapable conclusion is that for complete security the key must be random, must never repeat, and must therefore be as long as the complete sum of messages to be coded. However, even this statement needs a final clarification. It is true only in so long as the code contains at least an equal degree of complexity (the word 'entropy' is sometimes used) as the message. In essence, this requires that the key should possess at least as many different symbols in it as does the alphabet. Thus, if the key is a random selection of all 26 letters of the alphabet the message is secure. On the other hand, it is quite possible to set up a completely random and never repeating sequence of smaller 'entropy' than the message which would not be secure.

As an extreme example of this let us consider a completely random sequence of ones and twos such as

$$11212222212122212121111\ldots$$

actually obtained by me at my desk by tossing a penny (1 for heads, 2 for tails). If we interpret a 1 to mean 'replace this letter of the message by the next letter of the alphabet' (that is, A by B, B by C, etc) and a 2 to mean 'replace this letter of the message by the next-but-one in the alphabet' (that is, A by C, B by D, etc) then my original message concerning the red pencil box becomes encoded as

$$JICWGCTGFQGOEKNCQY.$$

If the possibility of a 'one and two shift' key were suspected, then I believe that you will easily convince yourself that decipherment can be obtained without great difficulty.

Random letter-for-letter substitution with a one-time full alphabetic key is therefore the solution to the unbreakable cipher. Because of its assured security, one-time keys of this kind have frequently been found on the person of detained foreign secret agents—one of the more celebrated being Colonel Rudolph Abel of the Soviet Union when he was apprehended in New York in 1957. It has also been reported that the 'hot-line' between Washington and Moscow uses this same kind of system (fully mechanized of course) with a one-time key.

In spite of their absolute security, however, these one-time key systems are obviously not at all suited for transmissions of vast quantities of classified information, such as would be required, for example, in times of war. In fact, during the Second World War the best systems for permanent transmission purposes used one-for-one letter substitution with a very long, but for practicality eventually repeating, key. Such well-known super-secret wartime codes as the German ENIGMA, British TYP EX, American SIGABA and Japanese RED and PURPLE machine codes were all of this kind. They were therefore all susceptible to 'breaking' by use of sophisticated statistical methods. This, perhaps, was the golden age of code breaking and the task was often quite a formidable one. Indeed, it took America well over one year of painstaking 'cryptoanalysis' by a veritable army of code breakers to finally 'crack' the Japanese code PURPLE in the summer of 1940.

As a general rule, codes are more difficult to break if they deny the would-be code-breaker as much statistical evidence as possible. Short of using the never ending and never repeating random key, some additional confusion can be added to letter-substitutional systems by substituting letters two or three at a time rather than singly. Although this can never completely remove the statistical evidence of language patterns, it does lend itself to the production of very simple keys and it certainly muddies the statistical waters for the potential code-breaker.

The simplest conceivable scheme of this kind sets up the letters of the alphabet in the form of a five-letter by five-letter square (with J omitted since it can be replaced in English by I without unduly confusing messages). The letters are placed within this square in some random fashion such as

```
T Z R M E
K O A Y P
F V D B N
U H G X S
C L W Q I
```

and pairs of letters are substituted by other pairs according to some simple rule. For example, if the two letters are not in the same row or column then the simple transformation pattern made clear by the examples of TO going to KZ, PR to EA, and RS to GE, would suffice. For letters in the same row one might perhaps use a rule like 'move up a row' so that, for example, GH goes to DV, and TE to CI (if we imagine the square key repeating itself like a wallpaper pattern with no 'edges'). Similarly, for letters in the same column, we might move them (say) one column to the right e.g., WA goes to QY, and PI goes to KC. Using this scheme my pencil box message

IHAVEAREDPENCILBOX

now transforms to the ciphered form

SLDOPRWIANTFUSVQHY

as can easily be verified by using the square key and the associated rules as set out above.

Methods of this kind can be made increasingly complex and correspondingly harder to break by inventing schemes for transposing letters three, four, or even five at a time. In this way, they can be made extremely difficult for the code-breaker to 'solve'. Their advantage is the unusually compact form of the key (if the methods for transposing the letters are committed to memory). Their major weakness is the comparatively difficult and time consuming nature of both the coding and decoding processes, together with the increasing probability of making errors as the procedure is made more and more complex in striving for added security.

All these systems of coding so far discussed, in spite of their varied forms and degrees of security, have one thing in common; they are *symmetric* in the sense that both the coder and the intended recipient have to be in possession of the relevant key to the code before any information can be transmitted. Now this is a great inconvenience and is also a very great threat to security since it means that, not only must one identify in advance *every* possible individual or organization to whom information might need to be supplied, but every one of them must also be provided with (and must therefore protect the secrecy of) the key. Should any single one of these intended receivers fall into 'enemy' hands with their key then the entire system breaks down and is rendered useless.

The general reaction to this, until very recently, has always been 'too bad, but that is the way it has to be!' Amazingly, however, such is not the case, and this astounding realization was first made in the mid-1970s. And if you have been wondering where numbers (either at work or at play) entered into the context of the present chapter, then the answer is 'right here' in the concept of an *asymmetric* code. The idea was first published in 1976 by research workers at Stanford University in California. They called it a 'public-key' coding system because, in using it (in the manner to be set out below), a person can announce to all the world the manner in which anyone wishing to send him a secret message should encode it. This could be done, for example, via a public catalogue. Now, incredibly, even though everyone can encode a message to him, only he is able to decipher it. Consequently, it becomes completely unnecessary for each potential pair of users to exchange, and guard with their lives, the same key in advance of their decision to communicate. Each receiver possesses his own decoding secret; he, and only he, needs to protect it.

Consider, for example, a military commander who has a number of patrols out scouting in enemy territory, each of which must be able to report its intelligence back to him in secrecy from the enemy. It now matters not one iota whether the enemy captures one of the encoding machines and hence discovers the encoding 'key' since, when they intercept other ciphers using the same code, this information will be of no help whatsoever in assisting them to decode the information. The system remains secure; only the commander possesses the decoding secret (or, in practice, the decoding machine).

In publishing the instructions for coding, to make them universally available, the potential receiver is deliberately giving half the cryptographic secret away. His motive for doing this is that then absolutely anyone (with the necessary ability) can code information to him, even people whose very existence he is completely unaware of, and yet he remains the only individual on earth with a knowledge of the other half of the key which is absolutely necessary to decode it. This all sounds a bit like black magic and since, in its actual machine application, it does involve a manipulation of extremely large numbers, it will be necessary for us to trivialize the procedure to see how it works. Instead of dealing with large numbers, the sheer size of which introduces the complexity that defeats the would-be code-breaker, we shall illustrate the technique using numbers small enough to manage with pencil and paper (or, at the very worst, with the help of a pocket calculator).

Think of a number made up by multiplying two prime numbers together. Let me choose for demonstration purposes the number 14, which is 2×7 (both primes). Now subtract 1 from each of these primes to give 1 and 6, and multiply the latter together. But 'Why?', you ask. Trust me ... trust me, for the moment. Let me symbolize this new number $1 \times 6 = 6$ by the Greek letter φ, since it plays a very special role in the coding and decoding systems as we shall see. I now pick a number which has no factor in common with φ. Since 2 and 3 are the factors of $\varphi = 6$, this leaves open to my choice *any* number which is not exactly divisible by 2 or 3. Let me choose 5 since it is the smallest such number available, and I want to keep my numbers as small as possible for my 'trivial' demonstration.

This number 5, together with my original number 14, are the ones which I publish to the world in my public catalogue. In order to code a message to me you are to proceed as follows. First replace the letters of the alphabet by numbers in the obvious fashion $A = 1, B = 2, C = 3$, and so on. The coding is then performed by replacing the number corresponding to any particular letter by that number raised to the fifth power (5 being the first of my catalogue numbers) but recording only the remainder when counting in groups of 14 (my second catalogue number). In the language of the last chapter, we count on a 14-clock, or (mod 14).

Let us see how this works by coding one very short and easy word BED. Firstly, translating it to numbers via the $A = 1, B = 2, C = 3, ...$ recipe, it becomes 254. It is now encoded by noting that

$$2^5 = \quad\; 32 = \quad\; 2 \times 14 + 4$$
$$5^5 = 3125 = 223 \times 14 + 3$$
$$4^5 = 1024 = \quad 73 \times 14 + 2$$

or equivalently, in clock language,

$$2^5 = 4 \qquad (\mathrm{mod}\ 14)$$
$$5^5 = 3 \qquad (\mathrm{mod}\ 14)$$
$$4^5 = 2 \qquad (\mathrm{mod}\ 14)$$

which presents us with the remainder sequence 432. It follows that 432 is the ciphered form of the word BED for this particular scheme. In order to decode it is necessary to have the 'key' which will turn 432 back into 254, or its alphabetic equivalent BED. What is this key? It is, in essence, a number known only to the receiver of the cipher. To obtain it requires a knowledge of φ. In detail, it is the number which, when multiplied by the first catalogue number (in our case 5), leaves a remainder of 1 when counting in groups of φ (in our case 6). This secret 'code breaking' number for our trivial example is therefore 11 since

$$5 \times 11 = 55 = 9 \times 6 + 1$$

or, in modular form,

$$5 \times 11 = 1 \qquad (\mathrm{mod}\ 6).$$

Decoding is now performed exactly as the coding was carried out, but using the secret number 11 as the new power to which the cipher numbers must be raised. Thus, taking the coded form 432 we proceed to calculate the remainders

$$4^{11} = 2 \qquad (\mathrm{mod}\ 14)$$
$$3^{11} = 5 \qquad (\mathrm{mod}\ 14)$$
$$2^{11} = 4 \qquad (\mathrm{mod}\ 14)$$

which, marvel of marvels, regenerate the original uncoded 254 or BED. Exactly why this particular prescription, involving φ, always gives a number which perfectly decodes in the manner set out above, is a question wrapped up in the mysteries of higher mathematics and clock numbers. But work it does, no matter how large are the numbers with which we are dealing. This is the essential point because unless we move to extremely large numbers (using, of course, a computer to do the coding and decoding) we have not really achieved anything new.

For example, the trivial exercise set out above is only a complicated manner of achieving a letter-for-letter substitution. In fact, it even fails in that because, since it counts (mod 14), it can have only 13 different remainders and there are 26 letters of the alphabet, even if we do not code punctuation marks, spaces, and the like. Also, it does not distinguish between 254 = BED and another possibility 254 = YD (Y being the 25th letter of the alphabet). This last problem is easily overcome by using two digits for each letter; i.e., A =

01, B = 02, C = 03, ... , X = 24, Y = 25, Z = 26. In this system BED has the numerical form 020504 while YD now appears quite different as 2504. This new system also has the added advantage that all the numbers from 27 up to 99 remain open for other uses such as representing more exotic symbols such as !, @, ♯, ∗, (,), $, %, &, ?, and many others.

The enormous power of the asymmetric coding system appears in two separate ways when large numbers are involved. Suppose, for example, that I chose as my starting number (taking the place of 14 in the trivial example) a 60-digit number formed by multiplying together two 30-digit primes. I am now able to code 30 consecutive letters and symbols (two digits per symbol) at a time. The code is therefore no longer equivalent to a one-for-one (letter) substitution, but to a 30-for-30 substitution of an extremely unusual kind. For example, the change of a single letter, say letter number 14, in the original first 30 'letters' of text, may well alter all 30 two-digit 'characters' in the coded form of this same set. The code is therefore essentially unbreakable unless the secret decoding number (11 in my trivial example) can be deduced by the would-be code-breaker. I can derive it since I know the exact identity of the two original prime numbers involved. I can therefore calculate φ, and from it choose coding and decoding exponents exactly as was done in the trivial example. The code-breaker can do likewise only if he or she can factorize my 60-digit number into its two 30-digit prime factors.

Now the whole secret of success for the asymmetric coding system as described in the preceding paragraph is that the problem of testing 30-digit numbers for primeness is (for a modern-day computer) a comparatively easy one, while the problem of factorizing a 60-digit number into two 30-digit primes is an extremely hard one. I can therefore publish my 60-digit number as part of my 'public catalogue' information with reasonable confidence that, even though everyone knows what it is, no-one will be able to factorize it and 'break' the code. However, spurred on precisely by this challenge to break asymmetric codings, factorization methods are improving by leaps and bounds at the present time (as will be discussed in the following chapter). As a result, 60-digit coding numbers have already become very 'shaky' from a security standpoint, and numbers up to and exceeding 100 digits in length are now preferred.

This discovery of secure asymmetric coding schemes seems likely, in the near future, to completely revolutionize the science of cryptography. Problems do remain, perhaps the most serious of which is the complexity of the coding and decoding processes which, at the present time at least, use up far greater computer time than equivalent amounts of comparably secure codings of the symmetric variety. If this difference is inescapable, then it may mean that both the symmetric and asymmetric coding systems will continue to exist and be used according to the requirements of each particular application. Neither, as used in practice, is unbreakable (in spite of the newspaper banner headlines asserting just that when the discovery of asymmetric coding was first announced to the public in the late 1970s). Just as the symmetric substitutional methods become unbreakable only in the limit of having a never repeating random key of sufficient 'entropy', so the asymmetric scheme becomes unbreakable only in the academic limit of using numbers of infinite length. The essential advantage of the latter is that the full key need be held only by the potential receiver. It is therefore exceptionally secure from a physical (as opposed to a technological) point of view.

9

Numbers and National Security

What are the prime factors of $2^{193} - 1$ and who cares? Factors, you will recall, are those numbers which, when multiplied together, make up the number in question. With this definition the answer to the first part of our question is 13 821 503, 61 654 440 233 248 340 616 559 and the even larger 14 732 265 321 145 317 331 353 282 383, prime numbers all. The answer to the second part of the question is, incredibly, the United States Government. 'You must be joking', I hear you say; but this, most certainly, is not a joking matter. The topic in question is computer security and the rapidly growing problems of computer fraud, theft and spying.

Large computer systems need an adequate level of protection and, since such computers have not been around for too long, company executives and government officials are only just beginning to understand how to provide protection. In the past it was obvious to all that sensitive papers were to be locked up. It is not so easy to grasp how to secure electrons flowing in wires, or pulses of light travelling through hair-thin glass fibres. Computer systems have progressed rapidly from the era of large metal cabinets and whirling magnetic tapes all collected in one air-conditioned room (which could be secured by simply locking the door) to systems involving terminals which may be scattered throughout a building, a city, or even around the world.

These networks are vulnerable in surprising and unexpected ways. Physical protection of the system is no longer possible, and focus switches to safeguards written into the computer programs themselves—for example, the use of secret codes such as those discussed in the last chapter. Associated situations may involve the transfer of funds between banks (major banks, for example, electronically process tens of *billions* of dollars a day for their customers) or may even involve national security. Couple this with the fact that there are more than 100 000 computer sites in the United States and Europe that are constantly talking to one another—transferring funds and transmitting critical data—and the extent of the problem becomes apparent. Some estimates of the losses attributable to computer-assisted crime already run into the billions of dollars per year. And what has all this got to do with factorizing numbers? Those of you who survived the ramblings of the last chapter to the bitter end already know. For the rest (shame on you!) I will briefly explain.

About a decade ago, when the problem of computer security was first becoming apparent, a protective system was proposed that involved a coding scheme based on very large numbers which are made up of the product, that is, multiplication, of two large prime numbers. By this we mean that the code can only be broken if, given the product number, its two prime factors can be determined. The entire idea is based on the fact that, while large prime numbers are comparatively easy to determine by computer calculation (and are even easier to multiply together), the reverse problem of deducing large prime factors of large numbers is extremely difficult. For example, it would only take a few minutes, using the most modern computer algorithms, to locate two 40-digit prime numbers and multiply them together to form a number with about 80 digits. At the present time, the reverse problem of factorizing this 80-digit 'composite' number (by anyone other than the persons who carefully prepared it) is at the very limits of research capabilities. It follows that a coding system based on such a number would today be fairly secure. But for how long? Factoring research, once one of the very loneliest of mathematical backwaters, is now a veritable hot-bed of activity due almost entirely to its association with coding and computer security. We shall therefore reflect a little on numbers and their factors.

Every whole number is made up out of prime numbers in a very special way such that each, in essence, has a unique personality. This occurs because every whole number can be formed by multiplying together one, and only one, particular set of prime numbers. As an example, the number 60 is 2 times 2 times 3 times 5. These numbers, 2, 2, 3 and 5, are called the prime factors of 60. Given the number 60, these factors can be found by testing each prime number in turn in increasing order (2, 3, 5, 7, 11, ...) for divisibility without remainder until the smallest 'factor' is found, dividing by this factor, and continuing the process to larger prime values. Evidently, as the number gets larger it becomes more tedious to factor, although the method, in principle at least, is valid for any finite integer. If no factors have been found when all primes up to the square root of the number have been tested, then the number can have no factors other than itself and one. Accordingly, it must be a prime number.

It follows that all numbers of comparable size (say with 10 digits) are by no means equally hard to factor. Those with many small prime factors will decompose into their ultimate fully factorized form much more easily than those which have no small factors. The latter kind are referred to as 'hard numbers' to factorize as opposed to 'easy numbers' of the same digit length. The brute force method of factorizing set out above soon becomes impractical for 'hard numbers' as digit length increases. Thus, for example, that hard number referred to at the start of this chapter has 58 digits when written out in its full decimal form. Even using the fastest of modern-day computers which can perform, say, one division every nanosecond (or one *billionth* of a second) it would still take more than 35 000 years of computer time to find the factors shown by that method. Clearly, the factors were not obtained by that method. They were in fact obtained in 1983 in a little less than 100 minutes by a group of workers at the Sandia National Laboratories in Albuquerque, New Mexico, using a powerful computer, but with a much faster method of factorizing.

This new method was the most recent breakthrough in factoring discovered by the Dutch mathematician Hendrik Lenstra. The method surprised and delighted the mathematical community both by its simplicity and ingenuity. Nevertheless, it has one unfortunate (or fortunate if you are a cryptographer) weakness in that it works efficiently only for cases which possess prime factors of significantly different sizes. For numbers whose prime factors are approximately equal in size, the new approach is only about as fast as the best previously known methods. Lenstra's breakthrough is of immense importance to number 'purists', since most numbers do not fall into the latter category. However, the large numbers used in asymmetric cryptography are, as discussed in the last chapter, purposely 'cooked up' to have just two prime factors of comparable size. They therefore still remain safe from the latest factoring breakthrough. But for how long?

In the short term Lenstra's method will probably divide the factoring community into two parts; the purists and the code-breakers. Purists factor numbers simply because they are there. They are, it has been said, like stamp collectors who try to fill in missing gaps in their number collection. They build up lists of 'wanted' and 'most-wanted' numbers, examples which have long resisted factorization, and gradually 'knock them off' one by one. The code-breakers are more militant. Supported by the National Security Agency and the Defense Department, they have no special number favourites but are out to bring any number to its knees. The question they confront is 'is factoring necessarily a 'difficult' problem?'. Lenstra's method, which at least makes most factoring much easier than it was once considered, is really the first effort to make use of 20th century mathematics for factoring purposes. Previous methods used ideas from the 18th century (primarily those of Gauss). Could it be that Lenstra's efforts represent only a new beginning—a first focus of mathematical 'big-guns' on this until recently relatively neglected pursuit? Maybe the really big breakthrough in factoring has yet to come. Until someone actually proves that the factoring of 'hard numbers' is necessarily hard in some absolute sense, you have to wonder. If Federal money can crack the problem then, presumably, cracked it will be. Maybe 10 or 20 years from now people will no longer be talking about factoring because it will be easy to do. If so, then neither will they be talking about asymmetric cryptography since its security will have disappeared.

In the light of all this glamour and news-focus on the number factorers of today, it is now only fair to go back and give a little credit to those earlier academics and pencil pushers who struggled with the problems of factoring long before it was profitable to do so. Let us go right back to the beginning. There have always been mathematicians and amateurs interested in factorizing even though, until the last few years, it was primarily the domain of the proven eccentric. The earliest tables of factors were constructed in association with research on prime numbers. Such a table for all the numbers up to 24 000 existed as long ago as the early 17th century and, in the year 1668, it was extended to 100 000. Publication of these results was not easy since, not surprisingly, there was little interest (let alone demand) for such works among the population at large. In this context perhaps the most ill-fated table was that painstakingly assembled by the Viennese schoolmaster Antonio Felkel who, in 1776, with the assistance of a distinguished German mathematician of the day, managed to persuade the Austrian Imperial Treasury to finance its publication. It contained factorization decompositions for all numbers up to over 400 000. However, there were so few subscribers to the volume that the Treasury shortly recalled almost the entire edition and converted the paper into cartridges to be used for making war against the Turks.

In the 19th century, mathematicians made use of a human lightning mental calculator named Z Dase to extend the factor table to 10 000 000. Such lightning calculators appear in history from time to time and normally have no great aptitude for formal mathematics. They possess the one very specific and exceptional talent associated with rapid numerical calculation, usually achieving these mental feats by methods as unknown to themselves as to anyone else. The pinnacle of factorizing achievement of the 19th century was the work of a certain J P Kulik, a professor of mathematics at the University of Prague. His unpublished manuscript, the result of a more than 20 year hobby, covered *all* numbers up to 100 000 000. Unfortunately (and perhaps not surprisingly), it is now known to contain more than a few errors.

Nothing approaching this magnitude of effort has ever, to my knowledge, been actually published. Also, as our interest expands to numbers of 20, 30, 40 digits, or even longer, it is clear that no such table is feasible even in principle. Interest is rather transferred to the problem of factoring any one particular integer, not in listing the factors of all numbers up to some limit. As the reader will have grasped by this time, there is no simple efficient means known for achieving such a task. Many methods have been developed over the centuries for factoring certain specific kinds of numbers but none is general. One can test the number in question by the various methods and hope—or nowadays, of course, one can turn to a machine for help.

The earliest such machine which met with impressive success was constructed by Dr D H Lehmer of the University of California, and his father. It was assembled in the early 1930s (long before the invention of the transistor which ushered in the age of the electronic computer) and was wholly mechanical. Inside the machine were sets of identical disc gears each having

100 teeth and mounted rigidly on a common drive shaft. Each of these gears enmeshed at either end of a diameter to other disc gears with different prime numbers of teeth. Each of the 'prime number' gears had holes in them, one for each tooth, and a beam of light could shine through these holes when, and only when, they were lined up. Although the theory behind the exact operation of the device cannot be detailed here, the basic set-up was arranged such that a simultaneous alignment of all the holes (signified by a beam of light passing through and falling on a photodetector) located a successful factorization.

The thrill of stepping into unknown number realms to search for divisors has been set out by the inventors themselves in an article entitled 'Hunting Big Game in the Theory of Numbers' in *Scripta Mathematica* **1** (1933). 'It would have surprised you', Dr Lehmer Senior said, 'to see the excitement in the group of professors and their wives, as they gathered around a table in the laboratory to discuss, over coffee, the mysterious and uncanny powers of this curious machine. It was agreed that it would be unsportsmanlike to use it on small numbers such as could be handled by factor tables and the like, but to reserve it for numbers which lurk as it were, in other galaxies than ours, outside the range of ordinary telescopes.'

The first task given to the machine 'in the outer galaxies of numbers' was to find the factors of the great unconquered factor 1 537 228 672 093 301 419 of $2^{93} + 1$. This number was already known (by methods of the kind discussed in chapter 7) to be composite, but it had no small divisors within the reach of factor tables. In about three seconds, as Dr Lehmer recalls the event, the 'eye' of the machine at the side of the whirling wheels gave the signal and the machine came to a stop. Upon examination it was found that this 19-digit number had been decomposed into two prime factors 529 510 939 and 2 903 110 321. It was a breathtaking moment—success in three seconds on a problem which had defied solution for years. Excitement was indeed rampant in those early heady days of adventure into the unknown. Many unclimbed numerical mountains were surveyed and many finally conquered. Little did they know how relatively feeble these efforts would appear only a decade or two later when the real power of electronic computing was brought to bear on the problems of factoring.

The modern era of factoring large numbers on digital computers really begins in 1971 when a famous 40-digit number succumbed. It may seem strange to the amateur that any large number should be termed 'famous', but mathematicians know which numbers are hard to factor and, at any given moment, they always have a short list of 'most-wanted numbers'. These are not just long numbers with no obviously small factors, but they are often numbers whose factors are important to engineers or pure mathematicians for doing specialized algebraic tasks. About this same time, the American Mathematical Society decided to sponsor mathematicians in their search for factors of large numbers. The idea was to make a table of all the factors of numbers of the form $a^n + 1$ and $a^n - 1$ where a is a small whole number and n

is a large integer. These numbers have always been of particular importance in number theory and are also used by engineers to generate random numbers. The American Mathematical Society Table was to be called the Cunningham Project Table in memory of a British colonel who, around the turn of the century, compiled a partial table of numbers of this kind. With this inducement, a concerted attack on these 'Cunningham numbers' was mounted during the 1970s, and methods of attack were refined. Numbers up to 50 digits seemed possibly to be within reach, dwarfing the 20-digit specimens which yielded to Lehmer's bicycle-sprocket and chain contraption.

Although today's computers are extremely fast they are still not, as we have seen, capable of finding factors of very large numbers by brute force methods. Special tricks have to be devised and, basically, what mathematicians end up doing is taking the problem of factoring one large number and breaking it down into related problems of factoring thousands of smaller numbers. Not all these smaller numbers have to be factored completely; it may only, for example, be necessary to know if particular prime numbers do or do not divide them exactly. But each of these smaller problems provides some information about the original factorization and together they can be combined to achieve the ultimate objective.

By constantly refining the computer process, the pursuers of factors had, by 1982, reached what they considered to be the effective limit of the method using the fastest computer then available (the Cray 1). This limit brought within range hard numbers of approximately 50 digits. Unfortunately, the 1982 list of 'most-wanted numbers' contained several of 60 digits or more and the prognosis seemed 'guarded' at best. Each autumn since about 1970 a group of mathematicians interested in the factorization problem had met in Winnipeg, Canada, to discuss progress, and the 1982 meeting therefore seemed destined to close-off the subject as far as the present generation of computers was concerned. It was agreed to go to press with a paper which summarized the situation and, interestingly enough, Dr D H Lehmer (inventor of the original mechanical factorizing machine of the 1930s) was still involved to the extent of being co-author. As one of the attendees at that Winnipeg meeting has said 'you could have collected money from anyone there if you bet that a 60-digit hard number would be factored in the next year'.

What suddenly changed the picture was not faster computers but a developing interest in examining the details of exactly how the computer hardware performed its numerical tasks. The breakthrough occurred over a beer at the Winnipeg meeting. After a day at the conference, the factoring group just happened to be discussing the major difficulties associated with time limitations on computers in general, and on the fast Cray computer in particular, when a Cray Research Engineer joined them. This fortuitous coming together of the minds soon had everyone buzzing with excitement because it transpired that the specific repetitious tasks which needed to be performed to attack factoring problems were ideally suited to the architecture of the Cray. Many thousands of very long numbers had to be modified many thousands of times, but only at a small and very specific number of places each time. It so turned out that the Cray design enabled these changes to be made in a time proportional to the number of changes and not to the lengths of the numbers.

This news sent the factoring buffs rushing back to their Crays and, sure enough, what the engineer had said was true. The advantages were immediate and obvious. Numbers with 52-, 53-, and 54-digit lengths were successfully factored in a few hours. These were numbers which had beaten earlier efforts, some of which had used in excess of 100 hours of computer time. Then, fresh from these successes, they took on a 58-digit one from the very top of the 'most-wanted' list. It was that $2^{193} - 1$ number cited at the very beginning of the chapter. And to the delight of all, after about eight hours, it was beaten and its prime factors discovered. However, since each three-digit increase in number length tended to more than double the time needed to factor the hardest numbers by this method, they were again approaching a time barrier. It looked as though 60 digits might be close to a practical limit. But then came further technical improvements and, using them, they refactored the old 58-digit favourite $2^{193} - 1$ in less than two hours—a further fivefold improvement in speed. Shortly thereafter a 60-digit number was conquered, and then another 'most-wanted' number, this time with 63 digits, in a little over five hours.

Since 1983, improvements in technique have continued apace. The competition is both between the algorithms used to attack the factoring problem itself, and between the computers which are best adapted to efficiently carry out these algorithms. Significantly, one very new computer which belongs to NASA (and is to be primarily used for analysing satellite data) seems to be ideally suited for factoring, possessing both the right architecture and the right kinds of languages. It is currently being programmed for factoring and the expectation is that it will routinely be able to do even the hardest 70-digit numbers in less than one hour. All but the very hardest will, using Lenstra's techniques, now fall in much shorter times.

Perhaps the ultimate effort is being made by a group of researchers who are actually building their own computer, which will specifically be designed to do factoring, and only factoring, in the most efficient manner which they can devise. The effort is being made as a joint effort of Purdue University and the University of Georgia. The machine is to be called EPOC (for extended precision operand computer) in official publications, but is more popularly referred to as the Georgia Cracker. The expectation is that most numbers up to 80 digits in length will be within reach of EPOC using times of less than one day.

However, the pinnacle of factoring achievement as I write (1989) has not been achieved by any single 'super-computer', but by the cooperative efforts of many. The latest trend has been to combine the calculations of large numbers of different computers, each working (usually during their off-peak night-time hours) on separate aspects of a single factorization problem. These then report back, via electronic mail, to an operations centre in order to coordinate their activities. The most successful use of this approach, using hundreds of computers in several different countries over the period of nearly a month, was recently reported in a triumphant first-breaking of the 100-digit barrier. Specifically, the number was the 100-digit 'hard' part of $11^{104}+1$ which remains after the 'easy' factor 11^8+1 has been removed. Its factorization into two (60- and 41-digit) primes was announced on 11 October 1988, at two o'clock in the morning, California time (milestone factorizations now being timed to the minute—somewhat like the birth of a child). Although this achievement could hardly be considered routine, it does suggest that any number much shorter than 100 digits may already be in danger. In fact, a successful 95-digit factorization was also accomplished in 1988 by means of personal computers alone—albeit using a veritable army of them over a three month period!

And where does all this leave national security? Well, the first asymmetric code developed for the government (called RSA after the last initials of its inventors) was based on the supposed inability of anyone to successfully factorize a hard 80-digit number. In the late 1970s, when the coding was being perfected, 80-digit numbers seemed very secure indeed. We now see that this code must already be a bit shaky. Consequently, it is now being proposed that newer codings be based on significantly larger numbers— possibly up to 200 digits. Surely these would be safe—at least for a while.

But is there necessarily a limit to the size of a number which can be factored? Is factoring necessarily hard? And even if it is, what about the foreseeable future? The limit, using present techniques, is essentially set by cost. As of today (1989) it has been estimated that it would take about ten million dollars to 'buy' the factorization of a 140-digit number in, say, one year. And what about that 200-digit number? By extrapolation, that would seem to come close to 100 billion dollars—an astronomical amount to be sure, but in truth only about half the annual interest on the national debt of the United States.

But there are more than mere secret coding systems which depend on the inability of people to factorize large numbers. Methods ensuring fairness in all manner of business transactions via computer have also been devised using these same ideas. The problem is easy to state. How do you keep the other fellow from cheating when each of you is sitting in front of a computer terminal? Suppose, for example, that you want to exchange important business information. Who is to give his information first, and what is to stop the other party from then refusing to honour the exchange deal? In short, is it possible for two people to exchange secrets without the help of a trusted third party? And, anyway, how trustworthy is the third party? Once again factoring has come to the rescue.

Let us set up the simplest possible situation involving fairness. I sit in front of my computer terminal and you in front of yours. We agree to a small wager on next week's football game, but favour the same team. Alright then, let us toss a coin to see who takes which team. But who tosses, and how does the other person know for sure which way the coin fell? Can we be trusted not to cheat? The problem is overcome by the following ingenious mathematical method invented in 1981 by Manuel Blum of the University of California and Michael Rabin of Harvard.

Just as was done for the asymmetric coding routine, I take two large prime numbers and multiply them together to obtain an even larger number which, due to its length (say 150 digits), is impossible to factorize. I transmit this number N to you and ask you to factor it. You cannot, of course; but you are able (that is, there are computer methods which enable you) to verify that the number given is not a prime and therefore really does have factors. You pick at random a number M which is less than $N/2$ and try it. Almost certainly it will not be a factor. You now square this randomly chosen number M and record (using the clock arithmetic of chapter 7) the remainder R left over when you count up to M^2 in groups of N. You send this remainder R, and only R, back to me. Now there is a method, using the original factors (which I, of course, know but you do not) for me to generate all the other numbers less than $N/2$ which lead to this particular remainder. Normally there are only two such numbers and therefore I find myself with two numbers, one of which is your arbitrarily chosen M, but I do not know which one. I now choose one of these and transmit it to you. This is the equivalent of my coin toss. In doing it I have exactly a fifty–fifty chance of transmitting back to you the number M which you originally chose. If I do, then you have no increase in information and still cannot factor my original number. You therefore lose. If, however, I choose the other number then, after transmission, you possess two different numbers connected with the same remainder. With this information a simple method exists which enables you now to factor my original number and you win.

If you think carefully about this process, there is a fifty–fifty chance of either 'player' winning and there is no chance whatsoever of cheating. If I give you a number which has no non-trivial factors (i.e., is prime) you can check it out. If I give you back a number which does not have the appropriate remainder R, then again you can discover my effort to cheat. This basic procedure, which is called an 'oblivious transfer', provides the basis for protecting both parties' interests in all kinds of telephone or computer transactions. The whole concept of oblivious transfer can now be expanded to allow certified mail, such as contracts, to be sent directly from one computer to another without the need for an intermediary. The sender automatically gets a receipt that also confirms the nature of the message. The essential procedure is to encode the message as a string of digits in several numbers which are sent for factoring. Running through the 'coin toss' routine for each of these numbers eventually (possibly after several unsuccessful 'tosses' for some of them) enables the receiver to decode the message. The sender's record of the 'coin toss' events is his receipt. If there were a dispute, a judge could determine from the record of the transaction that enough information was supplied to enable the receiver to decode the message, and therefore that the message must have been received. The procedure can even be embellished to cover the simultaneous signing of contracts at different locations. Although these ideas are all very new, the expectation is that they will become essential features of business transactions in future decades—unless, of course, the unthinkable happens and someone devises a simple routine for factorizing any number, no matter how long or 'hard'. Perhaps this is one case where mankind will benefit most from mathematical failure.

10

Are Four Colours Enough?

Few things are more associated with childhood than the activity of colouring a picture book. Adults just do not colour picture books. Stressing this point is the political joke concerning the candidate who, wishing to emphasize the naivety and immaturity of his opponent, remarked that rumour had it that a fire had recently destroyed his opponent's house and library 'causing the unfortunate loss of both his books, one of which he had not even coloured yet!' Assuredly, then, colouring at its most rudimentary level is not generally considered to require great intellect.

The only basic principle involved in colouring a picture from a picture book seems to be that any two areas which possess a common boundary line (and which we shall refer to as being adjacent) should have a different colour. With this as the only restriction, brightly coloured pictures involving every colour in the box can now be created with varying degrees of subtlety and aesthetic appeal. From this delight, only a mathematician could extract a problem. Nevertheless, after a while, one question does suggest itself to the more mathematical rather than artistic practitioner; what is the smallest number of different colours necessary to complete any conceivable picture?

A little experimentation quickly reveals that three colours are certainly not enough. On the other hand it does seem difficult to create a picture which needs as many as five. The answer, therefore, is probably four. In any case, I think you will agree, whatever is the correct answer it does not seem to be one of the world's weightier mathematical concerns. But how deceiving first impressions can be. This so-called 'four-colour problem' defeated the world's best mathematical minds for well over 100 years! Even today it still creates a distinctly uneasy feeling among many of the 'old school' of mathematicians because its proof, which finally appeared in 1976, requires a vast amount of machine computation. It can therefore only be verified by experts who have access to powerful computer electronics. What sort of mathematical proof is that? And why should anyone believe a computer anyway?

The controversy arises because this is the first time in which a computer has been involved in an *essential* fashion in an important mathematical proof. Of course, one can always hope that a much shorter and simpler proof is still 'out there' somewhere, just waiting to be discovered by a new generation of fertile mathematical minds. However, it is quite possible that no easier proof exists, in which case the four-colour problem has (among all its other claims to fame) given birth to a new kind of mathematical proof—one whose essential complexity *requires* the use of fast computing facilities. Unsettling though this may be to those of us who sense something of a 'loss of innocence' in all of this, such could well be the wave of the future.

In spite of the 'tour de force' aspects of its final solution, the four-colour problem had much humbler beginnings. It was in the year 1852 that a certain 21-year-old Francis Guthrie, who had recently been a student in London, wrote to his brother Frederick commenting that it seemed to him that every map could always be coloured with only four colours if adjacent countries were required to have different colours. By adjacent countries he implied those having a common border of finite length, excluding the case of contact at a point. Thus, for example, one can colour a chess board with only two colours, even though it possesses many points at which four different squares touch. Also, by countries he implied single areas completely surrounded by a border, making the rules exactly those of the colouring book situation with which we introduced the chapter.

Now Francis' brother was still at University College, London, and was attending the lectures of a certain Professor DeMorgan, a prominent mathematician of his time. It was to this learned gentleman that Frederick took his brother's comment, since he could not find any way to determine for himself whether the four-colour conjecture was true or false. It was certainly quite clear that at least four different colours were necessary because it was quite easy to draw a pattern of four areas (or countries) each of which was adjacent to the other three. An example is shown in figure 9 with colours labelled A, B, C, D. But how would one proceed from here?

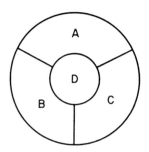

Figure 9

Professor DeMorgan took up the challenge and proceeded to investigate the possibility of having a situation in which five countries were each neighbours of (that is, adjacent to) the other four. If such a configuration could be drawn, then it would immediately be apparent that at least five colours were necessary. However, not only was DeMorgan unable to find an example of this type of connection, he actually succeeded in proving that no such set of five regions could possibly exist. This was a very considerable achievement and at first he was led to believe that with it the problem was solved. But unfortunately it is not, and in order to see why, let us go back to the case of figure 9 with four regions each in contact with three others.

Suppose that we draw a map which does not contain any regions like figure 9. Would such a map be colourable with only three colours? It is very easy to show that the answer is 'probably not!' Why not? Well, consider the map drawn in figure 10. It contains six countries, and within it there is no collection of four in which each is adjacent to the other three; and yet the map still requires four colours—three for the countries in the outer ring and one for the country in the centre. It follows that maps requiring four colours do not necessarily *have* to contain a configuration of four countries each adjacent to three others. By analogy, it may therefore equally well be true that maps requiring five colours do not necessarily have to contain that situation which Professor DeMorgan proved to be impossible—namely a group of five countries each adjacent to four others. The possibility that more than four colours were necessary for some maps consequently remained. The door was still ajar, but it is doubtful that any of the original participants in the drama would have believed that it would remain so for their lifetimes and some generations beyond.

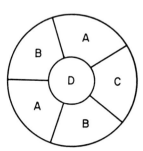

Figure 10

The story now moves on some twenty years or more to 1878 in which year the eminent British mathematician Arthur Cayley, unable to determine for himself the truth or falseness of the four-colour conjecture, decided to set the problem before the London Mathematical Society. Shortly thereafter, one of the members of that Society named Arthur Kempe (who was a lawyer by profession) published a paper claiming to prove that the conjecture was true.

Although the full details of this proof cannot be given here the steps involved are easily explained. Firstly, it was useful to introduce the concept of a 'normal' map. A map is called normal if no more than three countries meet at any point and if no country *completely* surrounds any other. In this sense (thinking in terms of states rather than countries) the map of the continental United States is not quite normal since the four states of Utah, Colorado, Arizona and New Mexico all meet at a single point. The country of Lesotho, which is completely surrounded by South Africa, is an example of the other kind of non-normal situation. Clearly, these situations are rarities in the real world of maps, and it is therefore comforting to find that they can immediately be excluded from consideration because it is very easy to modify any non-normal map to produce a closely related normal one which requires at least as many colours. If, for example, a map requiring five colours does exist, then a normal map requiring five colours also exists. We need therefore concern ourselves only with normal maps.

With this simplification Mr Kempe then proceeded to prove that every normal map must possess at least one country with five or fewer neighbours. He then showed that if any map requiring five colours had a country with fewer than six neighbours (which as he had just shown, it must have) then another five-colour map could be constructed from it which contained less countries overall. The argument then was the classic one known as 'reductio ad absurdum' or literally reducing the situation to an absurdity. For example, taking the new smaller five-colour map, one could repeat this operation to obtain yet another with still fewer countries ... and so on and so on *ad infinitum*. However, if in this way we start out with a non-infinite number of countries, then we must eventually reach a map with so few countries that five colours cannot possibly be needed. The argument is therefore reduced to an absurdity which implies that the original supposition, namely the existence of a map requiring five colours, must necessarily be false.

The problem therefore appeared to be settled and the 'proof' was accepted for a little over 11 years, until the year 1890 to be precise when, horror of horrors, an error was discovered and, even more distressingly, an error which did not seem at all easy to rectify. What had gone wrong? Well, Kempe's 'proof' consisted of two parts and the first, namely the demonstration that every normal map must contain at least one country with five or fewer neighbours, was sound. It was therefore in the 'reducing' part that the difficulty arose. The reducing argument is trivial for countries with only two or three neighbours (the one neighbour situation is equally trivial and has already been used to eliminate non-normal maps) so that we can easily follow it.

If there exists a map which requires five colours, then there must also exist a map of this kind with the fewest possible countries. Consider a country in this 'smallest' five-colour map which has only two neighbours, and centre all attention on this group of three countries. Suppose we remove the border with one of the two neighbours, thereby 'uniting' two countries into one. The new map has one fewer country than the smallest five-colour map and must therefore by definition be colourable with four colours. Let us so colour it and then replace the border. The recreated country is in contact with only two others whose colours are, say, colours A and B of the four choices available A, B, C, D. It may therefore be recoloured in either C or D, making the original map colourable using only four colours, in contradiction to the initial five-colour assumption. It follows that no smallest five-colour map can contain a country with two neighbours.

The same trivial argument can be carried out for a country with three neighbours, since we still always have one of the four colours A, B, C, D left to colour it with. The situation for a country with four neighbours is a little more difficult and requires the examination of more than just the immediate neighbours. In particular, it becomes necessary to show that it is always possible to avoid colouring the four adjacent neighbours with four *different* colours, in which case the fourth colour is always available to use for the centre country in question. Kempe, however, was able to produce a simple and extremely elegant proof of this which was also quite sound. It was in the final, and most involved, argument for a country with five neighbours, that an error was discovered in Kempe's work. The possibility of the existence of a five-colour map therefore still existed, even though no-one had actually been able to discover a specific example.

As the year 1890 drew to a close, the situation was therefore as follows. At least four colours were undoubtedly needed to colour a general map (or picture book). If a map needing five colours did exist, then the *smallest* such map had to contain at least one country with five neighbours, and none with less. The final step in the 'four-colour' proof could therefore focus on this country with five neighbours. Can such a country, together with its local environment, always be 'reduced' in the sense that one boundary (i.e., country) can be removed from the supposedly smallest five-colour map without decreasing the number of colours necessary to colour it? If it can, then no smallest five-colour map can exist, and four colours are indeed enough! Kempe's efforts in this regard, though flawed in some final details, are instructive.

The method of approach for the five-neighbour situation contains two parts. Firstly, an examination is made of the possible local configurations of countries which can exist in the general neighbourhood of a five-adjacent complex. From this one extracts what is called a 'complete set of unavoidable configurations', which is a set of local patterns of countries, one at least of which must occur in a real map. The second part is to establish that every single one of the patterns in this set can be reduced in the now familiar sense that one country can be eliminated without decreasing the number of colours necessary for colouring. Now there are an enormous number of quite separate unavoidable sets of configurations, some with few members, others with thousands of members. Kempe chose one with relatively few members and claimed that each was reducible. Unfortunately, one member could not be reduced in the manner which Kempe claimed—in fact it could not be reduced at all. On such a single weak link in an otherwise impressive chain of reasoning did the whole proof collapse.

In spite of Kempe's gallant failure, his approach to the problem remained by far the most promising available and was taken up again and again by many mathematicians over the next few decades. Most of the earlier work centred on reducibility; that is, of looking at various commonly occurring patterns and establishing whether or not they are reducible. In this manner a whole catalogue of reducible configurations was gradually built up. Those patterns for which no reducibility argument could be found had then to be avoided in choosing a set of unavoidable configurations. Some success was achieved for maps with not too many countries contained in them since, for these, the number of possible patterns (or configurations) is much smaller. By the year 1950 it had been established that every map with fewer than 36 countries could be coloured with only four different colours.

With respect to the general problem, however, things did not look good. Sets of unavoidable configurations with a manageable number of members always seemed to contain some which were irreducible. Sets which avoided these problem members always seemed to have many thousands of components, each of which would have to be proved reducible before the four-colour problem was solved. However, after 1950 a new ingredient was added; high-speed digital computers were becoming available and the possibility of adapting the problem for computer attack became attractive. The extreme pessimism of those earlier researchers whose efforts were restricted to hand computations now had to be re-evaluated.

Computer programs were soon developed to test for reducibility (at least for the more obvious forms of reducibility) and great progress was soon made in designating and cataloguing ever larger numbers of reducible configurations. In fact, although certain improvements in the attacks on reducibility have since been made, all the ideas on reducibility needed for the proof of the four-colour theorem were basically understood by the late 1960s. Unfortunately, comparable advances in the other phase of the operation—namely, the finding of unavoidable sets of configurations which did not contain any 'problem members' for which reducibility was doubted and certainly not proven—had not been made. Also it seemed as though configurations involving up to at least 18 countries would have to be included and, with configurations of this size, one ran into a computer-time bottle-neck. The problem was that the computer tests for (simple) reducibility took a period of time which increased fourfold for every country added to the configuration. Now although configurations of up to 11 or 12 countries could be handled at acceptable time and expense, this was close to the limit for the computers of the late 1960s. Indeed, it was estimated at the time that a single reducibility test for an 18-country pattern would take about 100 hours of computer time and much more storage than was currently available.

Further progress seemed to require a wait for the development of faster computers with larger memories. How much bigger and faster was not really known since no-one knew with any great assurance how many patterns would occur in a reducible unavoidable set, but it seemed likely that the number would be in the thousands. This was the situation when Kenneth Appel and Wolfgang Haken took up the problem at the University of Illinois in 1972. They began by looking for unavoidable sets which avoided members with more than a certain limiting number (say 15 or 16) of countries. These sets they called 'good configurations' in the sense that they might possibly, in the not too distant future, be testable for reducibility in a realistic time.

Over the next three years or so they continuously improved their techniques until, by the summer of 1975, they believed that they finally had a good chance of mounting a successful attack on the four-colour conjecture. From January to June 1976 the last details of the procedure were refined using the IBM 360 computer at the University of Illinois. Finally, in July 1976, the final triumph was announced. An unavoidable set of reducible configurations had at last been found. Over 1000 hours of time on three computers were used to finally pin it down, and the task required a demonstration of the reducibility of about 1500 configurations.

The programs and computer solution have since been checked by other groups and verified completely. Four colours do indeed suffice, as the proud Post Office of Urbana, Illinois, firmly stamped as part of its cancellation postmark on all the mail leaving that city in the days following the announcement of success. Why then do we still feel a bit uneasy about the whole thing? It is, of course, that none of us reading this or any other book can follow in detail or check out the proof for himself.

There is no denying that acceptance of this computerized proof involves certain acts of faith which are not necessary in more normal proofs. This is because, even if I read and understand every single line that the proof contains, I still have to believe that the computer is doing correctly the calculations which it is supposed to perform. So my belief in the proof of the four-colour conjecture is reduced to a belief that computers have been correctly programmed and carry out their calculations without error. I can test this (should I be an expert) only by appealing for assistance from another computer. It is therefore in some sense 'out of my hands'.

One mathematician has put it as follows: 'When I first heard that the four-colour theorem had been proved, my reaction was "Wonderful! How did they do it?". I expected some brilliant new insight, an idea whose beauty would transform my day. But when I learned the truth, I felt disheartened. My reaction was "So it goes to show that it wasn't a good problem after all" '. Unquestionably, a sense of disappointment remains. Even though we might accept the fact that the four-colour conjecture is true on computer evidence, there is a gnawing feeling that a simpler, more elegant proof must surely be around somewhere, just waiting to be discovered. That may well be the case, but unfortunately it is by no means certain. It is now known that mathematical problems which are simple to pose do not necessarily possess proofs which are simple to understand. In some cases, amazingly, proofs may not even exist *in principle*, even though the assertion stated in the problem is true.

The conclusion is that some simple conjectures may indeed be true, but either may not possess a proof at all, or if they do, may not have a simple proof in the normal sense. We now know that the four-colour problem does have a proof; we still do not know whether it has the sort of proof which can be verified by pencil and paper. But even if it does, maybe the time has come to recognize the possible existence of classes of problems which are better suited for attack by computer techniques. These would possess proofs which do not require the use of strong theoretical tools but have solutions of this new type, requiring individual steps which are very easy—but so many of them that the completion of the project is just not possible by pencil and paper in a human lifetime, or even in the lifetime of the universe.

There is, perhaps, an intuitive feeling that the verification of one computer's results by another's is still not satisfactory in some sense. But this argument is weak since, in calculations requiring excessive amounts of straightforward computation, computers are likely to be far more accurate than humans. A parallel argument, that all existing proofs (except the four-colour proof) are acceptably short and directly verifiable by humans, is also easily countered. After all, if you only use tools which are capable of providing short proofs, then that is all that you are likely to get. Other problems, accessible to computer attack alone, will simply be shelved in the 'unsolved' box forever.

So what, if anything, has come out of all of this effort? Quite obviously, whether or not a map requires four or more colours is of very little importance to map makers themselves. Has anything of serious mathematical consequence emerged? The answer, fortunately, is yes! One very important offshoot of the 120-year attack on the four-colour problem has been the development of a completely new branch of mathematics known as 'Graph Theory'. It concerns itself with ways of connecting paths between points in particularly efficient ways. How, for example, should we create airline routes or telephone lines which will serve the most people in the most efficient (that is, least expensive and most convenient) manner? What is the most efficient rerouting within the system if a problem should develop from overloading or breakdown? Such questions are of enormous importance in today's transportation and communications industries. In so far as progress has been made along these lines to make our lives more enjoyable, the four-colour problem has made a significant contribution.

However, over and above this, Drs Appel and Haken, who finally computer-solved the problem, hope that it will lead to an eventual cooperation, rather than confrontation, between the traditional methods of pure mathematics and the developing techniques of machine computation. Historically, they say, there is a precedent. From the time of the early Greeks right up until the Middle Ages, mathematics was considered to be a superior science to physics. It, after all, was built upon the firm foundations of logic and simple 'self-evident truths', while physics required messy experimentation and involved less-than-precise answers containing errors of measurement. Such an attitude set back the development of experimental physics for 2000 years until the arrival of Galileo Galilei (1564–1642) who, with a single measurement, disproved the erroneous assertions of Aristotle concerning the free fall of bodies under the influence of gravitation. These assertions of Aristotle were the result of pure thought processes alone and, on the strength of the assumed superiority of such abstract reasoning, had remained unchallenged for 20 centuries.

As soon as the importance of experimentation was accepted, the two disciplines (mathematics and experimental physics) achieved far more via their interaction than either could have achieved alone. Appel and Haken feel that their computerized solution to the four-colour problem may help to highlight the limitations of traditional approaches to mathematical proofs, and hope that it will lead to a successful cooperation in the future between man and computer within the framework of pure mathematics.

An interesting postscript to all of this concerns the fact that the four-colour restriction refers only to maps drawn on a plane or sphere (or surface which can be derived from either of these in a continuous deformation). For other fundamentally more complicated surfaces containing handles or twists, more colours are needed. Way back in the 1890s, Percy Heawood (who had first pointed out the error in Kempe's 'proof' of the four-colour conjecture) investigated some of these doughnut-like generalizations of the original problem. If a number n of non-intersecting tunnels are bored through a sphere, a surface of 'genus n' is said to be formed. The sphere is therefore a surface of genus 0, and a doughnut or a lifebelt a surface of genus 1. Surprisingly, Heawood showed that the colouring problem for these genus 1, 2, 3, 4, ... surfaces was simpler than that of the original four-colour problem, and he gave the general solutions for these seemingly more complicated situations.

Thus, the number of different colours required to colour any map on a doughnut surface was known some 80 years before the equivalent answer for a flat surface or sphere. For the record the answer is seven. Examples have been given in which all seven colours are needed, although they are by no means easy to invent. One of these sleepless nights you may like to inflate an inner tube and try your luck. One way is to construct a pattern in which each region of a given colour touches six other regions of different colours. On the other hand, as was demonstrated earlier in this chapter, it may be possible to stumble upon a more subtle arrangement (analogous to the four-colour pattern of figure 10). The real enthusiast will naturally not be satisfied until he or she has located examples of both types. It is certainly an interesting way to get to know your inner tube better.

Amazingly, except for the genus 1 (doughnut) surface, Heawood's proofs were also incomplete, although all his answers turned out to be correct. In fact, he presented a general formula which gave the largest number $g(n)$ of colours needed to colour any map on a genus n surface, valid for any n-value larger than zero. Thus, for example, $g(1) = 7$, $g(2) = 8$, $g(3) = 9$, ... (and no!, the sequence does not continue forever in the simple manner suggested by the first three terms). Heawood's oversight was that, although he proved that no map of genus n would ever need *more* than $g(n)$ colours, he did not establish (except for the doughnut) that less than $g(n)$ colours might not sometimes suffice. For example, it is very easy to prove that no more than five colours are ever necessary to colour a map on a plane or sphere but this does not establish the five-colour theorem for conventional maps. Happily, it turned out that Heawood's numbers were all correct although the final proof did not appear until 1969. Finally, it is interesting to note that, by sheer good fortune, if $n = 0$ is substituted in Heawood's $g(n)$ formula, it also correctly gives the four-colour answer $g(0) = 4$. Heawood, however, was never under any illusion that his method could be applied to that seemingly simplest geometry of all—genus zero—which gave birth to the infamous four-colour problem.

11

Rulers, Ominoes, and Professor Golomb

Few objects are more familiar than a one foot ruler. It enables us to measure and, in particular, if we focus attention (only) on its one inch markings, then it allows us to measure any integer distance between one and 12 inches. There is nothing very earth shattering about that, you may be saying to yourself; but have you ever considered the fact that this particular scheme is not a very efficient one? Why, if we label the 13 inch markings as 0 (say at the left-hand extreme of the ruler), through 1, 2, 3,.... all the way out to 12 at the right-hand extreme, then there are, for example, no less than seven different ways of measuring a length of six inches. We could, you see, use any of the combinations of inch markings (0,6); (1,7); (2,8); (3,9); (4,10); (5,11); or (6,12). Things are almost as bad for longer distances (there are four different ways of measuring a nine inch length) and are even worse for shorter ones. It is particularly easy to see that a single inch, for example, can be measured in no less than 12 ways.

Alright, you say, I accept that! But why should it bother me? I shall come to this point later in the chapter. But first, if you are willing to accept my promise of an eventual explanation, we can perhaps give some thought to the ways in which this general redundancy of measurement in all conventional rulers can be reduced, or even avoided altogether. As a general policy it is always wise to start such a project by thinking about the simplest possible cases. In the present context the ultimate in measuring simplicity is a ruler containing only two marks (which we label 0 and 1) just one inch apart. This is the 'trivial prototype ruler' which is, in spite of being of ridiculously little practical use, nevertheless perfectly efficient. By this I mean that it can measure its one possible distance in only one way.

The next step up in complexity is to think of a ruler with three markings. If we place them at positions 0, 1 and 2 (that is a two inch ruler with marks at each end and one in the middle), we have already generated an inefficient measuring device since it can measure a one inch distance in two different ways (0,1) and (1,2). This is a depressing situation so early in the game but do not despair, we are not yet defeated. Suppose that the three marks are placed instead in positions 0, 1, and 3. This is a three inch ruler with three marks on it and is rather clever since it can measure one inch (0,1), two inches (1,3) and three inches (0,3), but each in one way only. It is therefore perfectly efficient and is our first non-trivial example of a perfect ruler if, by the label 'perfect', we imply a ruler of general length N (say inches) which is capable of measuring all integer lengths from 1 up to N each in one way only. It follows that there are perfect rulers of length 1 and 3 but not 2.

A little bit of trial and error considering rulers with four marks soon establishes that there are no perfect rulers of length 4 or 5, but that the next-longer perfect ruler is one of length 6 with four marks at positions 0, 1, 4, 6. It can measure one inch (0,1), two inches (4,6), three inches (1,4), four inches (0,4), five inches (1,6) and six inches (0,6) and again each in only one way. This idea of perfect (i.e., perfectly efficient) rulers is an invention of Solomon Golomb, professor of mathematics at the University of Southern California. Surprisingly, they are not merely of academic interest. They have already been applied in coding theory, X-ray crystallography, radio astronomy and other fields in ways we shall expand upon when we have learned a little more about them.

The first further bit of information which we learn about them is seemingly the death knell of the entire subject, since a rather elegant proof has been given by Golomb that the three perfect rulers set out above are, in fact, the only three which exist. We may list them by their unit markings as follows

$$0, 1$$
$$0, 1, 3$$
$$0, 1, 4, 6.$$

A ruler with five marks can (if you think about it) measure 10 distances. If it were perfect it would therefore be of length 10, measuring 1, 2, 3, ... up to 10 units, each in one way only. However, for rulers with more than four marks, perfection is lost in one of two possible ways: either some distances can be measured in more than one way or some distances just cannot be measured at all. Given this rather unhappy situation, mathematicians, inspired by Professor Golomb's ideas, have forged ahead to look for the next best thing. For example, the next best thing to the non-existent perfect five-mark ruler might possibly be defined as one that contains each measurable distance only once, but which (unavoidably) cannot measure every possible distance up to the length of the ruler. But this alone is not a useful definition because there are countless numbers of this kind which can be invented at will. Consider, for example, the five-mark ruler 0, 4, 10, 27, 101; numbers which I have just picked out of the air. It can measure the distances 4 (0,4), 6 (4,10), 10 (0,10), 17 (10,27), 23 (4,27), 27 (0,27), 74 (27,101), 91 (10,101), 97 (4,101), and 101 (0,101), each one way only. What *is* a challenge is to find the very *shortest* ruler which contains five marks and is not 'redundant', by which we mean does not measure any one distance in more than one way.

The shortest five-mark ruler is in fact of length 11, just one unit longer than the 'hoped-for' perfect 10. It has marks at the positions

$$0, 1, 4, 9, 11$$

and can measure all distances up to 11 with the exception only of 6. This comes as a bit of a disappointment if we are not previously aware of Golomb's proof, since the search for a possible perfect length-10 five-mark ruler seems to be going well if we set off systematically in increasing lengths starting from the known perfect four-mark ruler of length 6. Thus, for example, we can find a five-mark length-7 ruler

$$0, 1, 3, 6, 7$$

which has a *three*fold 'redundancy' in the sense that it can measure three lengths (namely 1, 3, and 6) in each of two separate ways. Next we can find a five-mark length-8 ruler

$$0, 1, 4, 6, 8$$

which is only *two*fold redundant, measuring 2 and 4 in two separate ways, and then a five-mark length-9 ruler which is only *one*fold redundant. The latter has marks at

$$0, 1, 4, 7, 9$$

and, if you check it out, measures all the lengths between 1 and 9 in one way only with the exception of 3 which can be achieved in two ways (1,4) and (4,7).

The trend seems inescapable. The next step just has to be a five-mark perfect length-10 ruler with no degeneracy. But try as we will we cannot find it, because it simply does not exist. Oh! cruel world. In order to get a length-10 ruler which *will* measure all lengths from one to 10, it is necessary to introduce a sixth mark, e.g.,

$$0, 1, 3, 6, 8, 10$$

and the ruler is then embarrassingly redundant; in fact fivefold redundant since it can measure 3, 5 and 7 in two ways each, and 2 in no less than three ways. This seems incredibly wasteful. Numbers really should be better behaved than this. But swallowing this bitter pill of experience we can now march on to six-mark rulers. First we find one which is fourfold redundant, e.g.,

$$0, 1, 4, 7, 9, 11$$

of length 11, followed by a threefold redundant one

$$0, 1, 4, 7, 10, 12$$

of length 12.

Again we seem to be marching encouragingly towards a perfect six-mark length-15 ruler with no degeneracy, but once again it does not exist. Note, however, that at this stage we *have* answered at least one question of practical interest. It is 'how many of the 13 marks on a one foot ruler (marked in inches) can be erased without affecting its ability to measure all integer lengths between 1 and 12 inches?' The answer is that we can remove seven of them, leaving just the six marks, for example, shown in the pattern 0, 1, 4, 7, 10, 12 set out above. This will measure 3 (inches) in three different ways: (1,4), (4,7), (7,10); 6 in two ways, (1,7) and (4,10), and all the others in one way only.

In general the shortest ruler with *n*-marks is called the '*n*-mark Golomb ruler' in honour of its inventor. The six-mark Golomb ruler proves to be of length 17, rather than the hoped-for perfect 15, and all the Golomb rulers with up to 15 marks are now known. Beyond 15 marks we enter the zone of active research for which the Golomb rulers are, at the time of writing, not yet known. In the listing below we show all the known Golomb rulers:

Number of marks	Golomb length
2	1
3	3
4	6
5	11
6	17
7	25
8	34
9	44
10	55
11	72
12	85
13	106
14	127
15	151

For the record the largest known (15-mark) Golomb ruler has marks at

0, 6, 7, 15, 28, 40, 51, 75, 89, 92, 94, 121, 131, 147, 151.

The gradual increase in the length of Golomb rulers as the number of marks progresses is a fairly steady one although it has no easily recognizable pattern. For example, the increase in length in going from one Golomb ruler to the next larger generates the sequence

2, 3, 5, 6, 8, 9, 10, 11, 17, 13, 21, 21, 24, ?,

and already the 'smoothness' of the pattern is beginning to break up. That 13, for example, between the 17 and 21, would certainly not have been guessed in advance. And how long will the next discovered (16-mark) Golomb ruler be? Using the difference pattern one might guess at some value in the range 173–178. A 16-mark ruler of length 179 is already known which has no redundancy. It might just possibly be the next 'Golomb' but computers have not yet settled the question for sure.

Moving into the 'research zone', we list below the shortest known (1989) non-redundant rulers for rulers with between 16 and 24 marks.

Number of marks	Shortest known	Lower bound
16	179	154
17	199	177
18	216	201
19	246	227
20	283	254
21	333	283
22	358	314
23	372	346
24	425	380

Included in this table is a column labelled 'Lower bound'. This results from a known formula which locates the shortest length which a Golomb ruler with any particular number of marks can possibly have. For the 16-mark ruler it is 154 so that, as of this writing, the 16-mark Golomb ruler can possibly be of any length between 154 and 179.

We note that, in the research zone, the gaps between the shortest non-redundant rulers known and the lower bound (when expressed as the ratio of these two numbers) vary between limits of about 1.075 and 1.177. If these ratios are any indication, the 'best' estimates (lowest ratios) in the zone of research are for 18-mark and 23-mark rulers, while the worst (highest ratio) is for the 21-mark ruler. Although, by the time you read this book, the research will almost certainly have progressed further, there is opportunity for the amateur here. In cooperation with a personal computer it is quite likely that the enthusiast can improve on some of the 'shortest' rulers in the above table, although a demonstration that the actual Golomb ruler has been located is probably beyond all but the most powerful of today's computers. Unless, of course, someone comes up with a method of 'Golomb testing' more efficient than the present pedestrian procedure of simply generating and checking out every possible candidate.

Although all of this is quite fun as a computational exercise it is, believe it or not, also of practical scientific use. Not that usefulness has traditionally been a requirement for probing interesting properties of numbers. For the true 'number buff', numbers at play are every bit as intriguing as numbers at work. To this extent, the practical significance is mere 'icing on the cake'. One of the more recent applications has been in the field of radio astronomy, a science which studies the long-wavelength electromagnetic radiation coming to Earth from distant 'radio sources' way out in space. These sources are so far away that, to terrestrial observers, they are the very best approximation in the sky to 'point sources' or, if you like, objects with no measurable area. If only scientists can measure with extreme accuracy the direction in space from which such a radiation is coming, then it can be used to compute earth lengths (like the distance between the top of the Empire State Building in New York and the top of the Eiffel Tower in Paris) to an accuracy of inches.

The radiation, when it reaches Earth from its travels over countless light years of distance in space, is in the form (as far as our measurements are concerned) of *exactly* parallel wavefronts; like parallel waves breaking at an angle to the shore. To measure the angle at which they arrive from space (with respect to ground level in the vicinity of the experiment) a number of antennas are placed in a straight line several miles long. The precise time at which each wave in the incoming signal arrives at each antenna is determined with the greatest possible accuracy. Let us suppose, in a simplistic way, that each antenna records the time that the crest of a wave arrives. The problem which confronts the scientists at two different antennas is 'was it the *same* crest which they each observed or different ones?' For two people with stop watches at different points along the beach, doing a cruder version of the experiment on a sea wave, the answer would be easy to give because they could literally see the entire wavefront all the way along. In the radio-wave experiment the situation is more difficult. It turns out that the answer can be obtained, however, by positioning several antennas at the 'marks' of a Golomb ruler. If the distance between one pair of antennas is the same as the distance between any other pair then the two pairs provide redundant information. It follows that the greatest accuracy of angle computation for any given number of antennas is obtained when they are placed at Golomb ruler positions.

The final result of all this expensive experimentation is the ability to measure distances on Earth, or angular orientations, with unprecedented accuracy. Thus, for example, such fundamental quantities as the Earth's diameter, the orientation of the axis about which it spins, and the length of a day etc can be measured at the respective accuracies of centimetres, fractions of a second of angular arc and microseconds of time. This in turn makes possible the measurement of annual, seasonal, and even meteorological variations in these quantities which, as of today, are mostly baffling and mysterious, but which may soon lead to great strides forward in the physical understanding of the world in which we live.

Professor Golomb, in his earlier days as a student at Harvard, was also responsible for another mathematical brain-teaser which still defies solution. It concerns the object which he defines as a 'polyomino'. Now a polyomino (which I think of as a general member of the 'omino' family) is a figure formed by joining equal sized squares together at their edges. It follows that the simplest example (other than the trivial 'monomino', which consists just of one isolated square) is the familiar domino, from which the entire 'omino' nomenclature was derived.

Now there is really only one way of forming a domino from two equal sized (we shall call them 'unit') squares since we are not concerned here about orientation in space. That is to say, that for us a 'north–south' domino and an 'east–west' one are, so far as *shape* is concerned, the same. However, when one comes to the tromino (formed from three squares, remember, and nothing to do with those more recently retailed triangular tri-ominoes) it can be made up in two distinctly different fashions as regards shape. They are shown in figure 11. One is a chain-like creature while the other is shaped like a capital L.

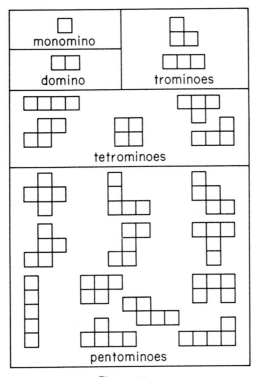

Figure 11

Quite obviously, this notion of dominoes and trominoes can be extended to ominoes made up of 4, 5, 6, 7,... unit squares of ever increasing complexity. The four-square versions, or tetrominoes, come in five distinct shapes, and the five-square versions, the pentominoes, in no less than 12 forms (see also figure 11). Thus, in attempting to answer the question of 'how many fundamentally different polyominoes are there with 1, 2, 3, 4, 5, ... unit-square components?' the answer begins with the series of numbers

$$1, 1, 2, 5, 12, \ldots .$$

But how does it continue? What is the formula which gives the answer for the general polyomino made up of n unit squares? Does such a formula even exist? Those were Professor Golomb's questions and, at the present writing, their answers are still not known.

On the other hand, since we now have access to large and powerful computing facilities, specific numerical answers for polyominoes up to at least 18 unit squares have been 'number-crunched' out of the machines. Up to $n =$ 18, the sequence proceeds as follows:

1, 1, 2, 5, 12, 35, 108, 369, 1285, 4655, 17073, 63600, 238591, 901971, 3426576, 13079255, 50107911, 192622052.

Mirror reflections are not counted, but shapes with interior holes are included. The ratios between successive terms seem to be settling down to the 3.8 to 3.9 region; in other words, each number near the end of the above sequence is a little less than four times the one before it. However, although formulas are known which give upper and lower bounds for the general problem, none is known which pins down the exact number.

The analogous problem is also unresolved for shapes made up when identical equilateral triangles are assembled into 'ominoes' by joining along their edges. You may like to find the first few sets of triangular polyominoes and see whether they lead to larger or smaller numbers than their square-unit relatives. The more ambitious of you can even go on to form hexagonal polyominoes made out of regular (that is equal-angled and equal-sided) six-sided units. All these problems remain unsolved in the general context. From a computational point of view they all appear to be 'hard' problems in the sense defined in the following chapter. However, to my knowledge, the question of whether they are truly NP complete (in the terminology of that chapter) remains uncertain.

12

What on Earth is an
NP **Problem?**

Dr Ronald Graham, of AT & T Bell Laboratories, once told a fascinating story about a fictitious bicycle assembly plant with a scheduling problem. It seems that (at least according to the story) a bicycle assembly consists of 10 separate operations such as frame preparation, wheel assembly, gear installation and the like. If we label these 10 jobs by letters (say A to J) then skilled assemblers were able to complete the various tasks according to the following table:

Task	A	B	C	D	E	F	G	H	I	J
Time	7	7	7	2	3	2	2	8	8	18

with the time in minutes. That rather lengthy job J was the final attachments, mountings and adjustments (involving handle bars, seat, brakes, etc).

Adding together the times as set out above, it would evidently take a single assembler 64 minutes to complete the job. Now it appears that this particular company had a bit of a space problem in the assembly plant to the extent that its 20 skilled assemblers just did not have the room to work independently each on a separate bicycle. They were therefore arranged into 10 teams of two workers, each team working together on one bicycle, with the expectation that each team would now be able to assemble a bicycle in half the time (that is 32 minutes) needed for a single worker to do the complete job. Looking at the table above, this could be accomplished, for example, by separating the job assignments in the following way:

Worker 1	Task	A	D	E	F	J
	Time	7	2	3	2	18
Worker 2	Task	B	C	G	H	I
	Time	7	7	2	8	8

There was unfortunately a difficulty. It is, you see, not possible to put together the separate parts of a bicycle in just any old sequence. Some jobs, like mounting the pedals, cannot be performed until others, like installing the gear assembly, are completed. It follows that some restrictions had to be recognized during assembly due simply to the practical realities of the job. In our lettering 'shorthand' it happens that these restrictions were as follows:

These jobs	could not be performed before	these jobs
J		A,B,C,D,E
C		D,E,A
E,F		D
H,I		E,F,G
G		F
B		A

Needless to say, this complicated matters considerably. In particular, it immediately ruled out the 32 minute plan given above. This is because the schedule requires job C (by worker 2) to begin after only jobs A and B have been completed whereas the table of restrictions says that C cannot be performed until jobs D, E and A have been finished. The work schedule obviously had to be rethought and, after a large amount of trial and error, the company eventually settled on the arrangement below for their 'standard':

Worker 1		A		B		I		H	
Time		7		7		8		8	
Worker 2		D	F	G	E		C		J	
Time		2	2	2	3		7		18	

which, if you check it out, does conform to all the practical restrictions but, unfortunately, takes 34 minutes, two longer than was originally hoped for. However, in spite of this, it did seem eminently acceptable since it also conformed to two other company rules—introduced generally to avoid time wasting—as follows.

Rule 1. No assembler can be idle if there is some job he or she can be doing.

Rule 2. Every job started must be completed as quickly as possible.

This procedure worked quite well for a while, until a backlog of orders began to pile up as the season of good cheer approached. It was at this moment that the company called in an efficiency expert to attempt to improve output by carrying out a time and motion study. After wasting numerous pieces of paper in an unsuccessful attempt to improve the scheduling, he opted to obtain the desired increase of output by renting all-electric tools. These tools, though expensive to rent, did enable every individual job to be performed one minute quicker than was possible with the old manual tools. Since there are 10 individual jobs to be done, and a minute can be saved on each, an improvement of about five minutes per two-worker team was anticipated. In other words it was hoped to crack the 30 minute barrier for single bicycle assembly—a target which the company had set for itself in order to meet the Christmas rush.

Starting off by retaining the old 34 minutes 'standard' job order, things did not go well. Some scheduling changes could not be avoided unless workers were to remain idle for periods of time when they could be occupied—a violation of work efficiency rule 1. Let us see what happened. Starting off on the 'standard' schedule, as set out above, a problem arose at the point

Worker 1		A	
Time		6	
Worker 2	|D|F|G| E	C?
Time	1 1 1 2

since worker 2 could not begin his standard schedule job C at the five minute point because worker 1 had not yet finished job A (and C cannot be done until A is complete). Since waiting for worker 1 to finish A would be a violation of efficiency rule 1, worker 2 had to find some other job which he could do, namely H or I. Obeying the efficiency rules to the letter then led amazingly to a two-worker single bicycle assembly time of 35 minutes—even longer than the time taken using the slower manual tools. Again, let us see precisely how this came about by setting out the exact work schedule:

Worker 1		A		B		H			
Time		6		6		7			
Worker 2	|D|F|G| E		I		C		J		
Time	1 1 1 2	7	6	17	

This is absolutely absurd, thought the efficiency expert. I obviously have to give a bit of thought to rearranging the work schedule to take proper advantage of the shorter job times. Unfortunately, try as he might, he could find no other schedule which could better the 35 minutes of that shown above. And no wonder, because none exists if the company 'efficiency rules' are adhered to.

In final desperation the rented tools were returned and the decision made to put more workers on the job. Fifty per cent more workers were hired which enabled three assemblers to work on each bicycle. Although the company recognized that this would not be cost effective in the long run, there was a desperate need to get out bicycles quickly over the Christmas period and it seemed time to panic. A little thought was obviously going to be necessary to arrange an efficient work order but, since the ideal situation should now approach a limit of 64/3 (or about 21) minutes per bicycle, even a relatively inefficient scheme should surely now be able to break the 30 minute barrier without too much trouble.

However, it was just at this moment that the real horrors began. Firstly, there is no way that all three workers can start the assembly in an active fashion since only two jobs, namely job A and job D, can be started without prior assembly. Newly hired worker 3 therefore is forced to start his assignment by sitting and watching. Not a very promising beginning! But worse was to follow. The best arrangement which the efficiency expert could find, using three workers and obeying all company efficiency regulations, was one requiring 32 minutes per assembly. Not only was this still outside the 30 minute target, but it was only two minutes better than the original 'standard' two-man schedule. The best three-man work arrangement obeying all the rules and restrictions was:

Worker 1		A		B		I				
Time		7		7		8				
Worker 2	D	E	G	C			J			
Time	2	3	2	7			18			
Worker 3		F			H					
Time	2	2	3		8		/			

Rumour has it that the efficiency expert was last seen running screaming from the premises. Had he known one other fact, he might well have been heading for the nearest cliff! You see, a two-man schedule exists which is better than the original 'standard' set out earlier. In fact, at 32 minutes, it is not only the equivalent of the three-man effort of the efficiency expert, but is perfectly efficient for the two-man team. Can you find it? Should I give it to you, or make you struggle to fully appreciate the difficulty of the task? I will relent (so that you can cheat if you wish!). The perfect solution is:

Worker 1		A		B			J			
Time		7		7			18			
Worker 2	D	E	F	C	G	I		H		
Time	2	3	2	7	2	8		8		

Although treated in a light-hearted vein this story has a very serious side to it. You see, some of the earliest motivation for studying this kind of scheduling problem arose from work on the design of computer programs for anti-ballistic missile defence systems. There it was discovered that decreasing job times, by increasing computer efficiency for individual component tasks, did not automatically make for a more time efficient complete project. For the military this realization was clearly a cause of some concern.

But going back to the bicycle company; what really had gone wrong? On the surface everything seemed to be geared for optimum performance. Actually, the villains were the 'efficiency rules'. They always imposed a short-sighted greediness without any regard for possibly serious problems which they were unwittingly creating further down the line. Workers were forced to start working on jobs which they were not allowed to interrupt when a more urgent job eventually came up. If only, in the rented-tools application, worker 2 at that critical five minute point had been allowed to twiddle his thumbs for one minute to allow job A to be completed. If he had, then the subsequent scheduling arrangement could have been very different; for example

Worker 1		A		B		I		H	
Time		6		6		7		7	
Worker 2	|D|F|G| E | |		C			J		
Time	1 1 1 2 1		6			17			

resulting in a joyous breaking of the 30 minute target (29 minutes to be precise). All the panic and final dementia of the efficiency expert would have been avoided, and the company's Christmas Season would have been just that little bit more financially jolly.

In a more general context, the difficulty is that it is often rather simple to program a computer to look for the most efficient next-step in a problem, but it is extremely difficult to program it to 'look ahead' or anticipate possible undesirable consequences of that short-term decision. Not all problems are plagued with unpleasant aspects of this kind. For some, a consecutive sequence of most efficient next-step decisions actually leads to the overall most efficient answer. Unfortunately, it now appears that a large fraction of the more important scheduling problems are of the other hair-tearing kind. These are, in some absolute sense, problems of a higher order of difficulty— and it is this particularly stubborn class of mathematical conundrums which has now come to be called NP problems. They were first identified as a group by computer scientists in the early 1970s and, for reasons far too complex to bother us, the NP actually stands for 'non-deterministic polynomial'.

These NP problems are not difficult in principle; only in practice. In principle, they can always be solved exactly by calculating (or more likely computing) every possible configuration and simply picking out the most efficient which satisfies the restrictions imposed (if any). What could possibly be simpler than that? The problem, of course, is that for all except the very simplest of NP problems, the number of different configurations is so unimaginably large. For them, even the most powerful of today's (and probably even tomorrow's) computers would take so much time as to make the obvious method of solution completely impracticable. In fact, many of the more important NP problems of interest in today's technological world would keep the computers busy for centuries (some for times even longer than the universe has existed) in order to locate with certainty the absolutely most efficient solution.

Not all of these NP problems are new. Some have been around for decades and they crop up in many practical situations. The important recognition, that these seemingly unrelated problems were all, in some mathematical sense, members of the same family, was made in 1971 by Stephen Cook of the University of Toronto. By this statement we mean that any efficient approximate method for finding a good (though not necessarily the best) solution for one of these problems is also an efficient approximate method for the others. Previously, mathematicians had been looking at each of these problems separately and searching for 'good' solutions to each as a separate task. After Cook's work the scientific community scrambled to see just how many of these previously unrelated scheduling problems could now be included in what was shortly to become known as the class of NP-complete (or simply NP) problems. At this writing several hundred have now been located and it would perhaps be fun to take a quick glance at a few of the most famous.

The NP problem which has perhaps received the most publicity of all is the travelling salesman problem. Suppose a travelling salesman wishes to visit 10 towns in a cross-country swing and then return home. It is obviously to his advantage, both in terms of time and petrol expense, if the route can be arranged to clock-up the smallest amount of mileage. From a purely mathematical point of view he has a choice of any of the 10 possible towns for his first stop, then nine further choices for his second stop, then eight and so on, making a total number of different possible schedules of $10! = 3\,628\,800$. However, since half of these routes are just reverse direction travelling from the other half, the more proper number is one-half times $10!$ or $1\,814\,400$. This problem already seems formidable enough from a formal standpoint. In practice, of course, the vast majority of these mathematically allowed possibilities could immediately be ruled out by a simple 'common sense' survey. In all probability our salesman would be able to deduce the ideal itinerary by simply testing a comparatively small number of the more likely looking schemes.

What if the number of cities was larger, say the 48 state capitals of the continental United States? The problem is now already an extremely difficult one requiring considerable computer time, although the exact solution has been known for many years. Suppose the salesman starts and returns to one particular capital, visiting 47 others en route. Without using any 'insight' at all this problem would then have one-half times $47!$ possible inequivalent routes, or approximately 1.3×10^{59}. Using a computer which can do one million such operations per second it would take about 4×10^{45} years to complete the task, or some 10^{36} times the age of the universe. Again, common sense obviously helps enormously and, in fact, the solution, first obtained in 1948, used a special method of linear programming which could obtain the *exact* best route for a salesman problem of this size, although the publication of this result was something of a landmark in travelling salesman history. But what if we are concerned with several hundred locations? The problem now, even with no additional restrictions, is already beyond the exact solution capabilities of modern-day computers.

Although it would seem hard on our travelling salesman to schedule him to visit several hundred cities, analogous problems with numbers of this magnitude are not difficult to imagine. For example, collecting money from coin telephones or vending machines, or even parking meters, might tax our optimum scheduling objective. At a more technical level one might consider the enormously more complex problem of routing telephone calls, which could today easily involve millions of calls between millions of locations, together with devising alternative routing arrangements when possible congestion or breakdowns occur. A better solution to this kind of NP problem can mean millions of dollars in savings to the telephone company.

Another hard or NP problem which has received considerable attention is the so-called bin packing problem. Suppose, for example, that there are a certain number of identical bins and a collection of rather odd-shaped packages. What, one might ask, is the smallest number of bins necessary in order to get all the packages in without any of the bins overflowing? Once again this basic problem is just one example of a whole range of related problems, which might include such more contemporary ones as the scheduling of television commercials of unequal length into one minute time slots, or the cutting up of standard-length boards to produce pieces of particular lengths for use in housing construction. Other examples are easy to envisage; for instance, those concerned with cargo loading, warehouse storage and budget planning.

The bin packing problems are ones which, in effect, turn the earlier scheduling problem around. Instead of asking how quickly a fixed number of workers can complete a given job subject to certain restrictions, we now ask questions which are mathematically akin to finding how few workers can be used and still complete a job by a given deadline. Translated into bin packing language the bins take the place of the workers, and packages play the role of component tasks which together make up the complete job. It follows that bin packing is also an NP problem and therefore extremely difficult to solve in the sense of finding the absolute best arrangement. It also follows that there are no known methods of exact solution except for the obvious, and usually impossibly time consuming one, of setting out (at least in principle) every possible configuration and looking at each in turn. If, for example, we are given 100 randomly chosen numbers less than a billion (10^9) which add up to a little less than 50 billion, then the problem of deciding whether or not they can possibly be packed into 10 bins of capacity five billion is at present beyond all the computing capacity in the world. This situation seems to strain credibility since we are only talking about 100 numbers (or packages) and there are certainly bin packing problems which go far beyond this degree of complexity. But then we were only talking about a few tens (or at most a few hundreds) of cities in our travelling salesman problem as well.

With the bin packing problem it is particularly simple to count the number of possible combinations which have to be tried to find out the answer. If, for example, we have only two bins A and B and two packages a and b, then the number of possible arrangements is four, or 2^2. We can put (1) a in A and b in B; (2) b in A and a in B; (3) a and b in A, nothing in B; and finally (4) a and b in B and nothing in A. If we now have three packages and two bins, a quick check locates eight or 2^3 combinations, and with four packages a total of 16 or 2^4 combinations. The pattern is suggestive and quite correct as it turns out. That is, there are 2^n ways of putting n packages into two bins. And if there are three bins this becomes 3^n and so on, so that the most general result is that the problem of packing n packages into N bins can be carried out in no less than N^n ways.

Since there are 'only' about 10^{75} atoms in the entire universe, it follows that the problem of packing 75 parcels into 10 bins has about as many possible solutions as there are atoms in the universe. The number problem set out above concerning the packing of 100 numbers into 10 bins therefore, with a now-known 10^{100} (which some call a googol) of possible solutions, has a complexity which can now more readily be appreciated. This googol is a number bigger than the number of atoms in a billion trillion universes. From what has been said above, concerning the capacity of computers, it is quite evident that no computer is ever going to be able to solve a problem like this (even the fastest and most ideally efficient futuristic computer that can be imagined) unless some fundamental mathematical breakthrough comes to its aid—and none is on the horizon at present.

What then is to be done? Do we just give up? Of course not! We merely humble ourselves a little and say that if we can find no viable method of obtaining the *absolute* best solution, then we can at least try to discover a much simpler method that will give an answer which is *close* to the best. What we need is called an 'algorithm', which is the word computer scientists use for a set of step-by-step instructions needed by the machine to tackle the problem. More precisely, what we need is a 'good algorithm', since we already have the 'bad algorithm' of computing every single possible solution and checking each of them out.

The difference between 'good' and 'bad' as far as the algorithms go is therefore something to do with time. In particular, an algorithm is bad if the computer time taken increases exponentially (that is, as some number to the *power n*) when we have a problem involving n packages. An algorithm is good if the time is related to the number of packages in a manner which does not involve an exponential increase of this sort. If, for example, the time involved was simply proportional to n (or even some power of n) then this time required for 'solution' would not 'explode' at such a phenomenal rate when n increased to modest values (say in the hundreds or thousands) as it does for the bad algorithms. Thus, for a good algorithm you can proceed *much* further up the n-ladder in a feasible amount of time (and at a tolerable expense) than would be possible with a bad algorithm. One can appreciate this by comparing two imagined algorithms, one with time varying as n^2 (i.e., a good algorithm according to our definition) and one varying as 2^n (a bad algorithm). The sense in which they are good and bad is clearly evident if we compare a bin packing kind of problem with 100 packages. For this case the bad algorithm requires 2^{100} divided by 100^2 times as long to complete the computer run. This works out to be more than 10^{26} times as long. Put another way, what the good algorithm could achieve in one second the bad algorithm would require more than a billion billion years to accomplish.

Quite obviously, 'good' algorithms are immensely important if only they can do an acceptable job in the sense of finding a 'schedule' or 'packing arrangement' sufficiently close to the absolute best. One of the most common of the good algorithms for constructing schedules is called the 'critical path' method. The basic idea is to try to choose, at each step, the 'most urgent' task to next start working on. By most urgent we mean the one that begins the chain of unexecuted tasks which has the longest time allocation remaining. These are the jobs which are most likely to create the bottle-necks further down the line. Such a criterion, in fact, gives the ultimate solution to the bicycle building example with which we started this whole discussion. On the other hand, it by no means assures us of an ultimate solution in every case; in fact, it can perform very poorly on some examples, so poorly that some scheduling problems can even be invented for which it gives absolutely the worst solution. In general, it works best when there are fewest time-ordering constraints among the tasks.

In the context of bin packing one of the simplest possible good algorithms is the so-called 'first-fit' method. In this we first arrange the items to be packed in order of decreasing size. We then fill bins by proceeding along the package line filling each bin as much as possible before proceeding to the next, and always returning to an earlier bin when an allowably small sized package is reached further down the line. As an example, we consider the task of fitting packages of 'size' 10, 9, 8, 8, 6, 6, 5, 5, 3 (of total size 60) into bins of capacity 20. The minimum number of bins is therefore three, although there is no guarantee that a solution requiring this ideal minimum exists.

Using the first-fit method we start by putting packages 10 and 9 into bin 1. The next package 8 is too much for bin 1 so that it goes into bin 2, as does the next package 8. The first 6-package will now have to go into bin 3 and so on. Proceeding down the package line we arrive at the 'solution' requiring four bins:

	3	5	
9	8	6	
10	8	6	5
Bin 1	Bin 2	Bin 3	Bin 4

Not bad! But does a three-bin solution exist? Yes, this time it does; it is

5	3	6
5	8	6
10	9	8
Bin 1	Bin 2	Bin 3

Does the first-fit method ever get the best possible arrangement? The answer is yes, and an example might be that of arranging the package sizes 14, 10, 8, 7, 6, 6, 4, 3, 2 (again adding up to 60) in the same three 20-capacity bins. This time the first-fit method leads to

		3
	2	4
6	8	6
14	10	7
Bin 1	Bin 2	Bin 3

which is a perfect packing. Generally, the method can do no worse than about 22% more bins than the optimum packing for problems involving large numbers of bins. Even so, some 'bicycle-like' frustration can confront the first-fit packing scheme in certain instances. For example, cases can be found for which the removal of some of the packages from the problem actually *increases* the number of bins required by use of this algorithm. These NP-complete problems just do not seem to like 'good' algorithms and conspire to frustrate them if at all possible.

Perhaps the greatest advance in computer algorithms for NP-complete problems was announced in 1984 by Narendra Karmarkar of AT & T Bell Laboratories in New Jersey. Although not leading to any exact solutions, it does reduce very markedly the time necessary for obtaining the best available approximate solutions. Moreover, its improvement over the previous best algorithm (the so-called simplex method, devised as long ago as 1947 by a mathematician named George Dantzig) becomes more and more impressive as the size of the problem increases. As a result, many problems can now be attacked which were simply 'beyond reach' before Karmarkar's breakthrough.

The startling degree of improvement has had an immediate application in the telephone business where this algorithm now helps to guide the routing of millions of calls between millions of locations. The associated problems were of a degree of complexity far beyond the capability of the old simplex algorithm, and the potential monetary saving to the telephone company is enormous. AT & T has filed for patents on various aspects of the applications of Karmarkar's algorithm and the achievement has already brought Karmarkar two major awards from the international scientific community. The method is still being 'polished' and will undoubtedly be improved as more and more of its details are refined.

13

How Many Balls Can you Shake into a Can?

Take a large known quantity of equal-sized balls and pour them into a cube-shaped can. Shake them up gently and measure the level in the container. Repeat the procedure and measure again, ..., and again. You quickly find that this level is remarkably stable if the shaking is performed in a thorough fashion, even though it seems inconceivable that the *exact* arrangement of balls in the can could be absolutely identical each time. Nevertheless, the effect is well known in real life; not only grocers, but their customers too, believe that the volume of a box containing one pound of coffee is well-defined. Although neither the grocer nor his customer may know how to calculate it, they feel sure that the first mathematician they meet on the street surely does. Their confidence in the abilities of those pursuing mathematical interests is commendable, but unfortunately it is misplaced. You see, no mathematician on earth knows how to do it either!

What we are looking for is the best way of packing spheres into a specified volume; we talk about finding the maximum 'packing fraction' which is defined as that fraction of the 'can' which is ultimately taken up by the balls. If this fraction is denoted by f, then the inevitable spaces between the balls must account for the rest of the volume fraction $1 - f$ such that the sum of the two parts (occupied and unoccupied) add up to one. What we are therefore saying is that no-one has yet been able to calculate the number f for the densest possible arrangement of spheres in three-dimensional space, a situation which we refer to as 'dense random packing'. Still more distressingly, we do not even have a convenient mathematical description of what random packing really is. Nevertheless, it seems quite clear from experiment that the number f exists, is reproducible with considerable accuracy, and is about $f = 0.64$.

But now suppose that I change the question a little. What if I do not merely shake up the balls, but place them carefully one by one in a manner which is

not random but possesses some repeating pattern? A few minutes experiment-
ing with golf balls or billiard balls will soon convince you that there is an
ordered dense packing which seems to be particularly efficient in three
dimensions. It is that often seen in fruit stands, or in piles of cannonballs at
war memorials, and is started by first arranging three spheres on a flat surface
with each one touching the other two. The centres then form an equilateral
triangle (which is one with equal sides and equal angles of sixty degrees) and
we continue the packing by adding spheres on the surface so that each new
one touches at least two of those already in place. In this manner we obtain a
layer of spheres in which each touches six others, except for those on the
'outside' where (for lack of time, balls, or patience) we decided to stop.

　　We now build a second layer of spheres on top of the first by placing them
in the indentations or depressions left at the centre of each triangle of spheres
in the first layer. The finished second layer is identical with the first, although
it is 'moved horizontally' in the sense that the centres of no two balls are
stacked vertically. More layers can now be added in the indentations of the
preceding ones until a complete three-dimensional ordered arrangement is
built up. The result is called the 'face-centred-cubic' packing and, by use of a
little bit of algebra and geometry, it is found to have a packing fraction f equal
to pi divided by the square root of 18, an irrational number which begins
0.740 48…. In this packing arrangement the balls therefore take up very
nearly 75% of the available space, and it is known that this is the densest (i.e.,
largest f-value) that can be achieved with an *ordered* packing, the proof being
first given by Gauss in 1831.

　　The improvement over that experimental value 0.64 for dense *random*
packing is impressive, and it is perhaps not surprising that careful planning
can do better than a mere shaking (although in our original problem the
boundary of the 'can' would in all probability prevent us from quite achieving
$f = 0.74$ for that case even with painstaking stacking). It is now very
tempting to assume that $f = 0.740 48…$ is the densest possible packing of
any kind in three dimensions. In fact it is often said that all scientists except
mathematicians know that this is true. Remarkably, however, from the
standpoint of rigorous mathematical proof the question remains open, since it
has never been established that some very cleverly designed packing which is
not ordered (that is, does not repeat itself in a regular manner) might not exist
with an even denser packing. This particular dense random packing would be
random in the sense that its pattern never repeated itself, but would also be a
very special packing in that it had an infinitesimal probability of arising by
chance (that is, by 'shaking'). The best that mathematicians have been able to
prove beyond doubt is that no packing of spheres in three dimensions can
have a packing factor f bigger than about 0.7796. However, this result is not
very helpful for anyone actually trying to build a more efficient disordered
packing, since the proof offers no clues as to how one should go about
preparing any arrangement which comes close to this upper bound.

　　Should we really take this possibility of a special random packing with f

larger than 0.740 48 seriously? That it is not a totally ridiculous notion is best envisaged by paying a little attention to what, in the present context, is affectionately referred to as the 'kissing number'. This kissing number is the number of identical spheres which can be arranged around a central sphere in such a way that all the surrounding ones just touch or 'kiss' the central one. In two dimensions we saw earlier, in arranging the packing of the first horizontal layer of the face-centred-cubic lattice, that this number was six. It is not necessary to have golf balls or billiard balls to answer the two-dimensional kissing question—pennies from your pocket will do. Put one down on the table and arrange others around it each to touch the original. In three dimensions, with a set of balls, the experiment is by no means so easy to carry out in a practical manner.

This three-dimensional sphere-kissing-number problem was, interestingly, the subject of a famous dispute in the year 1694 between Sir Isaac Newton and a Scottish astronomer named David Gregory. Newton maintained that the kissing number was 12 (which is the number found in the face-centred-cubic array) while Gregory believed that a 13th sphere could be squeezed in, although he was not able to prove it. The basic idea was that the 12 touching spheres from the face-centred-cubic packing could be rolled around the central one in such a way that the gaps, which are certainly there in this packing, could all be concentrated in one direction, thereby making room for an extra sphere to be inserted to touch the middle one. Only in the year 1874 was it finally established that the gap accumulated in this manner is not large enough to squeeze in the extra sphere, so that the correct answer to this particular kissing problem is Newton's, namely 12.

In spite of this finding, it is possible to start building a cluster with a few spheres in a way which seems to be doing better (from a packing density point of view) than the face-centred-cubic packing. Consider, for example, a 'seed' of four spheres packed together so that each touches the other three. When these four spheres are all touching, their centres are at the corners of what is known as a 'regular tetrahedron'. This is a figure with four identical triangular faces (i.e., it is a sort of triangular pyramid) each face of which has equal sides and angles. Since each sphere actually touches the other three, this must be the densest possible configuration for four spheres in three dimensions.

Now suppose that we add to this 'tetrahedral' seed other spheres, one at a time, so as to make up a new tetrahedral configuration at each stage. All that is required is for each ball to be added in such a way that it makes contact with three others. Take four billiard balls in contact and try adding the fifth; there seems to be no problem (although you might need a helping hand to actually set up the experiment). This looks great. If we continue in this manner then the packing certainly *will* have the maximum density possible; which is the upper limit $f = 0.7796$ mentioned earlier. What then is all the fuss about? The trouble is that as we proceed, there always comes a time when the next ball just cannot be added in the manner prescribed. At a certain stage of the 'growing' cluster, the surface acquires a shape which simply does not allow another sphere to be added that touches three existing ones. And when the breakdown comes it is disastrous, with large volumes of space now unavoidably being wasted in a manner which reduces the packing density dramatically to values below the face-centred-cubic 0.74, back towards the experimental limit of about 0.64.

What we have shown, in effect, is that three-dimensional space cannot be completely filled up with regular tetrahedra, in the way (for example) that it can be exactly packed with cubes. Filling space by fitting together similar shaped blocks is one of the oldest and most difficult of geometric problems, and the dilemma concerning the regular tetrahedra goes back over 2300 years to the days of Aristotle. He asserted (incorrectly as we now know) that, of the five regular solids which exist (these are solids with all side lengths and angles equal; see figure 12), the cube and the tetrahedron can be packed together to exactly fill all space.

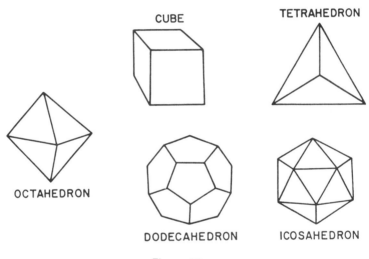

Figure 12

The discovery of the regular solids, and the proof that there are only five of them, was one of the great mathematical achievements of the ancient Greeks. They were discussed in detail by Euclid. Plato seems to have been the first to suggest that these solids might be the ultimate particles (or 'atoms') from which all matter is made up. Aristotle argued that Plato's idea was incompatible with reality since, of these five fundamental solid shapes, only two could pack together to successfully fill space without leaving gaps. A gap would mean an empty space which, according to Aristotle's ideas, could not exist in nature. Little did he realize that he had unwittingly made an errror affecting the number of balls that can be crammed into a can! Not for some 18 centuries after Aristotle made this mistake did the confusion become resolved. And even after the resolution, Aristotle's error persisted in various guises for a very long time, raising its ugly head to confuse many a space filling issue and embarrass many a geometer over the years.

So where does this leave us with regard to sphere packing? Quite evidently the 'greedy algorithm' of trying to make the next step the most efficient (without any regard for future difficulties) is somewhat analogous to, and shares the same fate as, the earlier efforts to solve the bicycle building problem of the last chapter. Mathematicians (and scientists, since dense random packings have physical consequences in the atomic arrangements of liquids and glasses) are still trying to come up with computer algorithms that will randomly pack spheres more efficiently than $f = 0.740\,48$ but, as yet, alas no success. It begins to seem extremely probable that nature has the correct answer after all since nature's $f = 0.64$ is persistently about the best result which can be obtained by even the most elaborate computer efforts to date. And yet the rigour of a mathematical proof is still missing. Physicists and chemists 'know' the answer; mathematicians, on the other hand, have to leave the door open just a crack.

But mathematicians, of course, being what they are, want to do much more than that. Why restrict your play to just one, two, or three dimensions? A world of higher dimensions is out there to be toyed with and to confuse us further, if only we can gear ourselves up to 'look' at it. Difficult though the kissing problem may be in three dimensions (that is, packing as many balls as possible around a centre one so that each touches the latter) it is doubtless going to be even more problematic in the fourth dimension, at least from a practical standpoint, since most of us have trouble enough merely wrestling with the existence of such a dimension, let alone a fifth, sixth, or even higher members of the dimension family.

In this respect mathematicians are much more fortunate than the rest of us since these higher dimensions (and in particular higher-dimensional spheres) are relatively easy to represent in a mathematical sense, even if they are tough to the common sense. For example, those of you who have done a little algebra and graph plotting in schooldays might recall that the equation for a circle is $x^2 + y^2 = 1$. By this we mean that if you take a piece of graph paper and plot y-values along one axis (say in a 'north' direction) and x-values along another at right angles to it (say 'east'), then those values of x and y which satisfy the above equation (like $x = 1$, $y = 0$; $x = \sqrt{(1/2)}$, $y = \sqrt{(1/2)}$; $x = 0$, $y = 1$ and countless other pairs) 'map out' a circle with a centre at $x = y = 0$ and a radius of 1. Try it out if you are not familiar with the notion. In three dimensions we need another axis to represent 'up'. If we label this as the z-axis, then the mathematical equation for a sphere with centre at $x = y = z = 0$ and radius of 1 is just $x^2 + y^2 + z^2 = 1$, a so-called 'unit' sphere. Since most of us do not have a three-dimensional piece of paper this is a little harder to verify, but rest assured that it is so. But, aha! We now begin to recognize a pattern. If we had a fourth dimension which we labelled w, then it would seem consistent to call the object defined by the equation $w^2 + x^2 + y^2 + z^2 = 1$ (if we could 'plot' it on a four-dimensional piece of paper) a 'four-dimensional unit sphere'.

As we go to higher dimensions we tend eventually to run out of letters, so that it is easier to label the 'axes' as directions $x_1, x_2, x_3, x_4, ..., x_n, ...,$ for which notation we can happily write down the equation for a 100-dimensional unit sphere if required, namely,

$$x_1^2 + x_2^2 + x_3^2 + x_4^2 + ... + x_{98}^2 + x_{99}^2 + x_{100}^2 = 1.$$

We may not know what to do with it, but that is not the point. The point is that the question of, for example, the kissing number in four, or even 100, dimensions is now just one of algebra and therefore has a precise answer which can be probed mathematically. Evidently, packing problems must therefore also have a mathematically well-defined representation and are consequently equally open to investigation. In this manner, we can break the bonds which lock us into a locally three-dimensional physical world, and a Euclidean universe of higher dimensions opens up for us to probe and ask questions about.

Being able to define a unit sphere, or ball, in an (arbitrarily large) n-dimensional space now enables us to answer questions like 'in what dimensional Euclidean space does a ball of unit radius have the largest volume or surface area?' Although the mathematics may be beyond most of us, the procedure is well-defined and the answers turn out to be $n = 5$ for volume and $n = 7$ for surface area. Amazingly, for larger dimensions the volumes get smaller and smaller, approaching zero as n progresses to infinity. The infinite-dimensional unit ball has no volume at all. Baseball or cricket would be a tough game in this limit!

The search for dense sphere packings in multi-dimensional spaces is made simplest if we focus first on 'lattice packings'; that is, packings which repeat in some regular fashion. The easiest packing of all is then one where each sphere is placed with a centre on each 'lattice point'. In two dimensions these lattice points might most simply be the points $x = m$, $y = n$, where m and n are integers, which make up the pattern of square centres on a checker board. The lattice is called D_2 and is shown in figure 13a. Its packing fraction f can be calculated to be 0.785 398... (pi over 4 to be exact), but we already know that it is not the densest lattice packing in two dimensions. The densest is the hexagonal packing of figure 13b which formed the base of our carefully stacked oranges or cannonballs. By elementary geometry its packing fraction can be shown to be close to 0.9069 ... and it is called a 'laminated lattice'; in this case the laminated lattice of dimension 2, symbolized L_2. The laminated lattice L_3 in three dimensions is built up by stacking layers L_2 in the third dimension in a fashion which 'nests' the succeeding layers in the deepest 'holes' of the layer below. This, as we have seen from our earlier orange stacking experience, creates the face-centred-cubic lattice. It is labelled L_3 and, as mentioned earlier, is known to be the densest *lattice* packing in three dimensions.

This procedure of building up laminated lattices can be mathematically extended to four dimensions and higher. For example, the lattice L_4 is formed by stacking L_3 lattices in the fourth dimension such that they nest together in the most efficient packing fashion. It is almost impossible to picture this because our simple minds are only able to visualize in the three-dimensional world of our physical experience. Mathematically, however, the procedure is quite precise and creates the sequence of laminated lattices L_4, L_5, L_6, ... and so on. Their properties have been studied in detail all the way up to 25 dimensions. Strangely, although there is one, and only one, laminated lattice L_n in all dimensions n up to 10, and again in dimensions between 14 and 24, there are two laminated lattices in dimension 11, three each in dimensions 12 and 13, and no less than 23 in dimension 25.

PACKING D$_2$ PACKING L$_2$

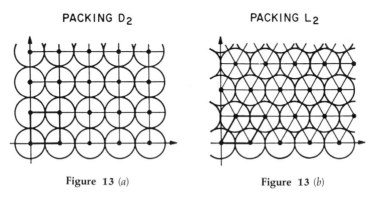

Figure 13 (a)   Figure 13 (b)

Out to the eighth dimension these laminated packings L_n have been proven to be the densest possible lattice packings, and in none of these dimensions has any denser amorphous (or non-lattice) packing yet been found. In fact, the laminated lattices are the densest known lattice packings for all dimensions out to 32 (examples of laminated lattices have, at this writing, been computed out to dimension 48) except for dimensions 10, 11, 12 and 13. Here, another sequence of lattice packings, called K packings, are known to be denser, while in dimensions larger than 32 yet another (so-called P packing) arrangement does better. However, in none of the dimensions greater than eight is the best lattice packing known for sure, and the question as to whether a non-lattice or amorphous packing is ever densest of all remains an open one. Interestingly, amorphous packings have been found in dimensions 11, 12 and 13 which are denser than any of the lattice packings yet found in those dimensions.

The size of the islands of packing knowledge in large dimensions compared with that of the ocean of ignorance can be gleaned by some recent work by N J A Sloane of Bell Laboratories who, in collaboration with other researchers in the field, has constructed mathematical packings in dimensions up to 100 000. In these uncharted waters the density of these new lattice packings is at least $10^{40\,000}$ greater than any simple multi-dimensional extensions of cubic packing, but still a factor of 10^{4000} smaller than are known (by general theorem) to be possible.

The dimension of 24 is rather a special case. For it, the laminated packing

L_{24} is closer in density to a theoretically established upper limit than for any other dimension larger than eight (for which the exact solution for densest lattice packing is known). This packing was first discovered by John Leech of the University of Glasgow in 1965. It is almost certainly the densest lattice packing which exists in 24 dimensions. Each sphere in this lattice touches 196 560 others, which is also the kissing number in 24 dimensions. This last result is remarkable since the kissing problem has only ever been solved for four other cases, namely the trivial ones of one or two dimensions (with kissing numbers 2 and 6 respectively), three dimensions (with kissing number 12 as was discussed earlier), and eight dimensions, for which the kissing number is 240.

Coming down from these lofty heights, our final degree of ignorance remains that we still do not know how many balls can be shaken into a can, either in three dimensions or in any dimension larger than three. Indeed, if we state the shape of the boundary of the can and substitute circles for spheres, there are many questions which remain unanswered even in two dimensions. You are all, I am sure, familiar with the process of 'racking up' 15 balls into a pool-table 'triangle' before the break shot which starts the game. This is just the problem of packing 15 circles into an equal-sided triangle of the smallest possible side length, and it is not difficult to prove that this particular 'pool-table' packing is the most efficient that can be achieved for this situation. If you now remove any one ball from the 'rack' you find that the other balls are still all tightly locked together in the sense that none can move. This would seem to establish that the smallest triangle which can pack 14 balls is exactly the same size as that which can pack 15 although, to my knowledge, no rigorous proof has been given. But what if a second ball is now removed from the rack? If you try it, you will find that the 13 remaining balls are certainly now free to move. The smallest triangle which can contain 13 balls is evidently smaller than that required for 14 or 15 balls. But how much smaller? And what pattern do the balls take up in the most efficient (that is dense) situation? It is amazing that even this very 'simple' situation involving only two dimensions and so few variables has not yet been solved.

The more general problem of packing identical circles, without overlap, into specified boundary shapes of smallest area, is one for which no general solution exists (even when the boundary region is as simple as a circle or a square). In each case the best possible packings have so far been located only for a rather small number of circles or for very special highly symmetric situations, like packing an exact square number (e.g., 4, 9, 16, 25, 36, ... etc) of circles into a square. Even in cases where exact answers are known the results are full of surprises. Thus, for example, for the case of packing circles into a circular boundary of smallest area, we show the best possible packings in figure 14 for two to 10 circles. We note that, for the cases with eight or nine circles, the densest packing contains a circle which is completely free to move. This is certainly not what intuition would suggest, but true it is! And does figure 15 look like the most efficient packing of 12 circles in a circle to you?

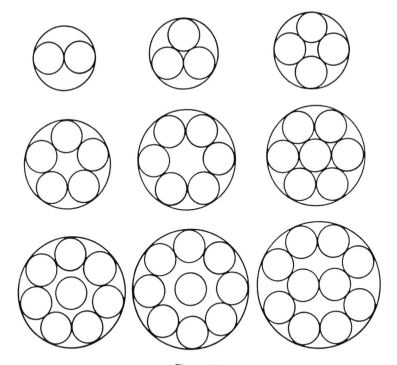

Figure 14

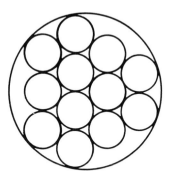

Figure 15

Well it is not. Can you think of a better one? There is at least one in which some circles only touch two of their neighbours, whereas in figure 15 each circle touches at least three neighbours—another seeming paradox.

The game, of course, can also be played in three dimensions in the form of spheres packed inside the spherical boundary of smallest volume. Even less is known about this problem, except for the very smallest number of spheres. Cube packings into cubes is even more difficult unless, once again, the numbers of cubes involved are exact cubic numbers (e.g., 8, 27, 64, 125, ...) or very close to them. By varying the shapes involved, these packing problems can be cast into a virtually endless set of mind boggling formats.

Returning again to two dimensions, for which most of the very few exact results have been firmly established, we show in figure 16 the most efficient packing of seven circles into a square. This one again is very hard to believe. In fact, I would imagine that the reader could win many a wager by offering large odds to friends (or even better, to enemies) that they cannot find a better way of packing these circles together. That unattached 'odd-one-out' just has to be able to squeeze in closer, surely. But rest assured, a better packing for seven circles cannot be found, so the money is yours. The proof was first given in 1965.

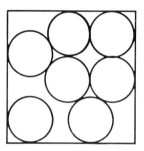

Figure 16

Finally, a mention must be made of the version of this packing game which involves packing squares within a square boundary. Suppose we want to pack as many as possible non-overlapping unit squares (that is, squares of side length 1) inside a square boundary of side length L. If L happens to be exactly equal to an integer (say N) then the answer is obvious, and a *complete* packing (with packing fraction $f = 1$) is trivially arranged. But what if, say, L is a little bit larger than N; maybe $L = N + (1/10)$? What should we do in this case? One obvious attempt at a solution is to fill up the N by N square which was the exact solution for $L = N$ and simply surrender the remaining narrow strip of uncovered area (of about $N/5$ square units) as unavoidable waste. Indeed, this does appear to be about the best one can do if N is a small number. In general, however, better arrangements can certainly be achieved, although exactly what the best formation is remains uncertain for most cases.

Such a sweeping statement can be made because of the existence of a proof that, as N becomes large, packings exist which leave an uncovered area considerably smaller than $N/5$. Although the minimum possible uncovered area is not known as a function of N, the existing proof establishes that the above 'N^2-plus-unavoidable-waste' arrangement is certainly not the best if N is greater than about 80, and is probably not the best for N-values as small as 30. On the other hand, exactly how we go about 'sqeezing-in' the extra squares is not at all obvious.

In order to grasp the scale of the problem for larger numbers, imagine 10^{10}

unit squares packed nicely into a large square boundary of side length $L = 10^5$. They fit exactly, and the packing fraction is accordingly $f = 1$. Now increase the side length of the large square boundary by just *one one-hundredth of a single unit*; that is, to $L = 10^5 + 0.01$, an amount small enough to go unnoticed you might think. Yet that 'unavoidable' and almost invisibly thin strip of waste area now amounts to no less than 2000 square units, and it can be established that a judicious shuffling about of the 10^{10} unit squares enables at least another 520 (and probably more) squares to be added. Once again, however, how one actually achieves this in an optimum fashion remains completely unknown.

14

In-between Dimensions

Until well into the 20th century, mathematics (particularly as it related to geometry) was still couched in the same idealistic concepts that had persisted since the days of Euclid. Curves were smooth (except possibly for an occasional sharp corner) and could be represented mathematically by functions which were in some classical sense 'well-behaved'. At each point one could ask about the slope of a curve or about its degree of curvature and get sensible answers. Such 'smooth' curves were accepted as being fully capable of describing virtually all features of the orderly world in which we were considered to live. They were called 'analytic', and were special in a sense which I can perhaps get across to the non-expert by the following story concerning anti-aircraft gunnery.

In the days before the heat seeking missile, the degree of success in shooting down aircraft with ground-based guns depended quite considerably on some knowledge of mathematics. The problem was that the target was a moving one, so that in order to hit it, it was necessary to aim at the point where the aircraft would be when the shell got there. Thus, if you aimed directly *at* the aircraft you always missed because, by the time the shell got there, the aircraft had moved on. Some correction had to be made. If the speed of the aircraft was known (or could be reliably estimated) then a 'correction' could be worked out to take account of this effect. So far so good; the misses now became 'near misses' instead of 'far misses', but misses none the less. What was now wrong?

The new problem was that, since these particular aircraft were still in the process of taking off, their velocity was not constant. In other words they were accelerating. But, in principle at least, this acceleration (which is the rate of change of velocity) could also be estimated and a further correction made. Now surely a hit could be made. But no! The acceleration was also not constant, and yet a further refinement was called for. All this was very frustrating for the gunners, who finally gave up the calculations—supposing none the less that, if some superhuman could continue this correction process on forever, a hit would finally be assured (at least in theory). In actuality, they need not have felt so badly about their computational limitations, because it happens that even the full infinite number of corrections would be no help either. For example, if the moment that the shell was fired coincided with the pilot's realization that he had forgotten his lunch and would have to return for it, then the miss would still occur.

The reason is simply that the spatial position of the aircraft at *any* future time is not determined even by a complete knowledge of its past motional history. But all smooth (that is analytic) curves in mathematics have this special property that an *exact* knowledge of part of the curve does precisely determine the rest of it. We therefore conclude that the aircraft position as a function of time is not analytic. Analytic functions are everywhere in mathematics (because they are 'easy' to write down and to calculate with) but are virtually nowhere in the real world. Nature is just not simple and orderly enough to fit such a description. Clouds are not spheres, mountains are not cones, lakes are not ellipses, lightning does not travel in a straight line. In fact, the closer we look at objects in nature, the more we realize that most of them lack smoothness in a very complete sense. This sense is that they seem to possess the same level of irregularity, on a smaller and smaller scale, the closer you come to them. Perhaps the most famous example of this was cited by Dr Benoit Mandelbrot, the famous French mathematician who pioneered the exploration of the rough edges between dimensions, in his classic paper in 1967 entitled 'How long is the coast of Britain?'.

If you imagine a space traveller approaching Earth in a direction towards the British Isles, then the closer he gets the more detail of the coastline becomes visible. Smaller bays and headlands that cannot be seen at all from a distance gradually become clear and then, on closer approach still, these bays and headlands can be seen to have structures of their own. On the scale of a yardstick the details of the rock formations would need to be measured, and on the centimetre scale, the positions of small pebbles would have to be noted. Finally, if you were intent on inducing complete craziness, you could attempt to measure around every grain of sand—assuming that you could stop the motion of the water.

If 'best' estimates were made of the coastal length of the British Isles, then a peculiar effect would be noted. It is that the smaller the scale of measurement used for the project the longer the estimate would be. This is embarrassing enough, but what is particularly alarming is the fact that, if an effort is made to find a limit towards which these measurements are converging as the ruler gets shorter and shorter, then this limit seems to increase without bound (or in other words is infinite). In this sense the coastline of Britain in the limit of infinite precision is boundless or, if you prefer, is longer than any distance which you care to name.

Now this observation had, in fact, been made by others before Dr Mandelbrot gave it his attention. Not only did the length of coastline increase without limit as the length (say L) of the 'unit of measure' decreased, but it was known to increase in a power-law fashion; that is as L^n, where n is a negative number. For example, if $n = -\frac{1}{2}$, this would mean that every time the unit of measure was halved the coastline would increase by $\frac{1}{2}$ to the power $-\frac{1}{2}$, a number (for those who remember the rules of algebra relating to fractions and powers) equal to the square root of 2. Those of you who rely more on the electronic calculator than algebra these days can push the y^x button with $y = 0.5$, $x = -0.5$ and read the equivalent result 1.4142.... Good estimates of this number n had already been made for many coastlines and land frontiers around the world, with n in the neighbourhood of -0.2, but different (and seemingly reproducible) for different countries.

If a coastline or frontier were smooth in the analytic sense we should expect n to be zero, in which case its length would converge on some nice finite value as the measuring rod L got smaller and smaller. This is exactly what one would expect for a one-dimensional 'line' like the perimeter of a circle. What Mandelbrot did was to write a relationship between the power n and the concept of dimension, which we shall symbolize by D, in the form $1 - D = n$. For one-dimensional 'Euclidean' lines, with $D = 1$, this conforms with the expectation $n = 0$. However, for frontiers and coastlines, with negative values of n, it led to values of D such as 1.23, 1.18 and the like and introduced for the first time the concept of fractional dimensions.

The formal concept of fractional dimension had been used in a purely mathematical context as far back as 1919 by philosopher, author and (later in life) mathematician Felix Hausdorff, but only for rather pathological objects conjured up in the tortured minds of a few maverick geometers. Mandelbrot's claim was that the world was filled with such manifestations and that virtually every real-world object truly existed in the hitherto unknown region 'in-between dimensions'. The number D, now called their fractal dimension, measures something quite precise about them and, in the limits $D = 1$ and $D = 2$, is quite consistent with the classical picture of one and two dimensions. Dr Mandelbrot coined the word 'fractals', from the Latin word fractus which means irregular, to describe all shapes with non-integer values of D. Although fractals are not restricted to the dimensional region between $D = 1$ and $D = 2$ we shall, for the sake of simplicity, consider this domain first. For the record, the fractal dimension of the West coast of Great Britain seems to be about 1.25.

The fractal dimension of the West coast of Great Britain is only approximately known since it requires experimental measurement. The most precise ideas concerning fractals are obtained by inventing simple geometric procedures which lead to fractals. Perhaps the simplest of all, and the one which is most frequently used to introduce the idea of fractals, is the so-called Koch snowflake. In order to form this intriguing shape we start with an equilateral triangle (figure 17 (a)). Each side is then marked off into three equal parts, and each centre part then removed and replaced by two sides of another smaller equilateral triangle for which the third side is the (now removed) centre span. The resulting shape is the 'Star of David' as shown in figure 17(b). The procedure has resulted in each side of the original triangle being replaced by a 'kinky' line which is $\frac{4}{3}$ times as long as the original. The new perimeter is therefore $\frac{4}{3}$ times the length of the old one while the new straight edge is only $\frac{1}{3}$ the length of the old.

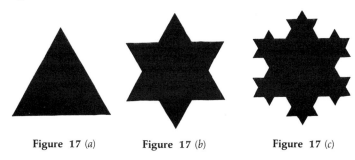

| **Figure 17** (a) | **Figure 17** (b) | **Figure 17** (c) |

If we call the original perimeter P_0 and edge length L_0, then the new perimeter $P_1 = \frac{4}{3}P_0$ and the new edge length $L_1 = \frac{1}{3}L_0$. If we now repeat the process using the new sides, we proceed to a second even more jagged shape (figure 17 (c)). Once again the length of the new edge is $\frac{1}{3}$ that of the old ($L_2 = \frac{1}{3}L_1$) while the perimeter is again increased by a factor of $\frac{4}{3}$ ($P_2 = \frac{4}{3}P_1$). Obviously, this procedure can be continued in principle *ad infinitum*, although the resulting 'snowflake' becomes more and more difficult to draw. After the fifth stage (e.g., P_5, L_5) the top side of the flake is shown in figure 18. Eventually, the 'spikes' become so numerous and so small that they just cannot be drawn by any means and, in the limit of progressing to infinity, there is ultimately a spike at every single point of the 'curve'. The curve has no continuous or smooth regions at all; it has become a 'fractal'.

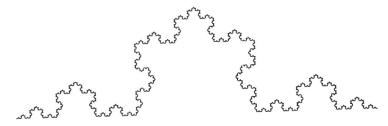

Figure 18

Although we cannot draw this fractal as it approaches its limiting form we do know something very special about it. It is that at each stage of the building process, no matter how far along we are (i.e., at the millionth, billionth or trillionth step), the perimeter P_{n+1} of the $(n+1)$th stage will always be $\frac{4}{3}$ times the perimeter P_n of the stage before it. In addition, the side length L_{n+1} of the $(n+1)$th stage will always be $\frac{1}{3}$ of the length L_n of the stage before it. We may therefore write the equations $P_{n+1} = \frac{4}{3}P_n$ and $L_{n+1} = \frac{1}{3}L_n$ for any value of n whatsoever. According to Mandelbrot's definition of dimension D we should also write P_n as a constant times $(L_n)^{1-D}$ and an equivalent equation for P_{n+1} as the same constant times $(L_{n+1})^{1-D}$. Dividing one of these last equations by the other we find the 'Mandelbrot equation'

$$(P_{n+1}/P_n) = (L_{n+1}/L_n)^{1-D}.$$

But since $P_{n+1}/P_n = \frac{4}{3}$ and $L_{n+1}/L_n = \frac{1}{3}$ for any n, it follows immediately from the above equation that

$$\frac{4}{3} = \frac{1}{3}^{1-D}.$$

The solution of this equation can be expressed in terms of logarithms in the form

$$D = \log(4)/\log(3)$$

which, for those of you who have a calculator with a log button on it, can be checked out numerically as $D = 1.261\,859....$. In comparison with the approximate estimate of 1.25 for the dimension of the West coast of Great Britain, the implication is that the Koch snowflake is just a tiny bit more 'jagged' than this particular coastline. The word jagged is, however, quite insufficient to express this difference since each is infinitely jagged in the normal sense of the word, and this is precisely where the idea of fractal dimension is so valuable. It measures a well-defined property of these infinitely jagged curves in a way that enables us to compare them in a quantitative manner.

The curve in figures 17 and 18, when continued to its infinitely jagged limit, was first explored by the German mathematician Helge van Koch (in 1904) as a precise example of a continuous curve of infinite length whose properties could not be described by the mathematics of the day. In the decade or so which followed, many other such curves were identified and discussed. However, the 'mathematics establishment' and mainstream academic community of that time regarded them with great suspicion, referring to them as a 'gallery of monsters', 'pathological' or even 'psychotic'. They were treated as diabolical constructions which had little, if any, relevance for the real world. It is now apparent that almost precisely the opposite is true, fractals seeming to be almost everywhere in nature; from leaves and flowers to river banks and lake edges, and from the scattered paths of nuclear particles to the awesome expanse of galactic clusters throughout the heavens.

The Koch snowflake is an example of what is called an *ordered* fractal in the sense that the rules for its construction are precise and exactly repetitive. Such ordered fractals have the property that any part of the curve continues to look *exactly* the same no matter how much we magnify it. We call it exactly self-similar. Coastlines and the like are obviously not quite so precisely self-similar. They possess this property in a statistical sense, and it is just this averaged self-similarity which enables us to define fractional dimensionality for such cases. We therefore refer to these less precisely ordered fractals as 'random fractals'. The implication is that self-similarity persists upon magnification in a manner which involves some averaged property (namely the fractional dimension) over and above the random changes in the detailed form.

Other ordered fractals are very easy to define and to construct in principle. In practice they are, of course, all impossible to draw, although computer graphic techniques are able to produce pictures covering many stages of their early repeated development. In order to produce curves of larger fractal dimension between $D = 1$ and $D = 2$ it is only necessary to define a more complicated manner of restructuring the line segment at each stage of the growth. We could, for example, start with a square and, instead of raising a triangle on the centre one-third of each edge, raise another square. At each stage of its growth this curve would increase the length of its perimeter by a factor $\frac{5}{3}$ and, if you follow an analogous calculation to that used for the snowflake example, arrive at a fractional dimensionality of $\log(5)/\log(3) = 1.464\,97\ldots$, or close to $\frac{3}{2}$.

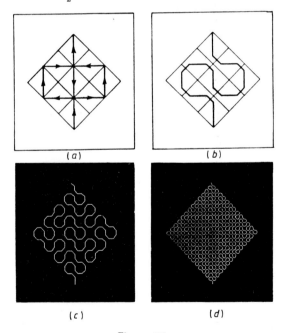

(a) (b)

(c) (d)

Figure 19

The question now arises as to whether we can study the gradual approach to $D = 2$ which, if the idea of a fractal dimension is to make any sense, should be a curve which fills up the whole of a two-dimensional area. Perhaps the most famous of such space filling curves was first proposed by the Italian mathematician Guiseppe Peano in 1890. The manner in which it is generated is shown in figure 19. The procedure starts with a square and a dividing of each of the four sides into three equal segments. In this way one can imagine the square divided into nine smaller squares as shown in figure 19(a). The Peano curve then starts by drawing a straight-line diagonal from bottom to top and then replacing it with a continuous line through the diagonals of all the nine smaller squares, following the pattern of the arrows in that figure. In order to help in 'seeing' the shape of this curve it is usual to chop off the corners of each right-angle twist as shown in figure 19(b). Neglecting this corner chopping in our formal development, we note that each of the nine small squares now has a straight-line diagonal across it and therefore looks exactly like the original square when we started. We can therefore separate each small square into nine even smaller ones and continue the building-up of the Peano fractal to one with $9^2 = 81$ (corner-chopped) line segments at the second building stage (figure 19(c)). This is then followed with $9^3 = 729$ line segments at the third stage (figure 19(d)) and so on *ad infinitum*.

It is, I think, fairly evident visually that this curve, when continued to the infinite degree of line fragmentation, will completely fill the area of the original square. It certainly looks like a $D = 2$ fractal. Is this in accord with Mandelbrot's definition of fractal dimension? Well, looking back to the first stages of the building process, the edge length L decreases by a factor of three at each stage while the perimeter length increases by a factor of three. Putting $P_{n+1}/P_n = 3$ and $L_{n+1}/L_n = \frac{1}{3}$ in the Mandelbrot equation now leads to the numerical relationship

$$3 = \tfrac{1}{3}^{1-D}$$

for dimension D. Even if you do not know much about logarithms it is only necessary to recall from chapter 1 the definition of a negative power or exponent (e.g., $3^{-x} = \frac{1}{3}^x$) in order to be able to verify the physically relevant solution $D = 2$ (i.e., $x = -1$). The formal definition is therefore consistent with our simpler concepts of integer dimensions. It is not difficult to invent other $D = 2$ fractals and, in honour of the inventor of the first such 'devil curve', all $D = 2$ fractals are now usually referred to collectively as Peano curves.

The concept of fractal dimension is by no means restricted to dimensions between $D = 1$ and $D = 2$. Particularly easy to imagine are fractals with dimensions smaller than one. These may easily be prepared in an ordered fashion. For example, if we start with a straight line and separate it into three equal parts, we can now consider removing the centre part completely. We are left with two lines each one-third the length of the original and with an equal length space between them. The new 'edge length' is $\frac{1}{3}$ times the original and the new 'perimeter' is $\frac{2}{3}$ times the original. Each of the two smaller line segments can now be split up in exactly the same fashion as was the original, and the procedure continued to infinity. We finish up with a sort of 'grey line' (if the original was black) made up of an infinite number of infinitesimally small line fragments. It is an odd looking object but, using

Mandelbrot's formula, we can easily calculate its dimension. Noting from the building procedure that $L_{n+1} = \frac{1}{3}L_n$ and $P_{n+1} = \frac{2}{3}P_n$ at every step n, it follows that

$$\frac{2}{3} = \frac{1}{3}^{1-D}$$

an equation for which the solution is $D = \log(2)/\log(3) = 0.6309...$. Obviously, similar 'cut-out' schemes can be invented using different details and leading to all sorts of other fractals with D between zero and one. In a sense the less grey the final fractal line becomes the smaller is the dimension until, as D approaches zero, the line disappears from view. An interesting physical model of this kind is provided by the structure of trees. The continued branching process, in which any branch splits into two smaller ones, and these smaller ones split again, and so on and so on, eventually generates a tree outline which Mandelbrot refers to as a 'fractal canopy' of dimension $D = 0.6309...$.

The concept of fractal dimensions can also be pursued to values greater than two. A perfectly flat sheet of aluminium foil, for example, is an excellent physical approximation to a two-dimensional surface. If, however, you wrinkle it up, you cause it to deform into a fractal of two-plus-some-fraction dimensions. The precise value of this fractal dimension depends on just how wrinkly the crumpled aluminium foil becomes and can be established (at least in principle) in a method exactly analogous to that used for the fractal snowflake curve. The only difference is that now we are concerned with progressive measures of area as we continually decrease the size of a 'measuring plate', which now takes the place of our earlier measuring rod.

As an ordered fractal of dimension 'two and a bit' we could, for example, start with a square-shaped area and divide it equally into nine smaller squares as was done for the Peano curve. This time, however, we consider raising a 'cube' from the centre square by 'pulling it out' in the third dimension. The object so formed now looks like a square hat for a man with a square head, and contains 13 equal-sized small square areas—nine horizontal and four vertical. Each of these may now be used as starting squares for the second stage of the building process, which raises smaller 'hats' on each. The procedure can obviously be continued to infinity where, since the surface area increases by the fraction $\frac{13}{9}$ at each stage of development, the final surface area is infinite, indicating a fractal with dimension larger than two. I leave it to you to verify that the actual value of the fractional dimension of this crinkly object is $\log(13)/\log(3) = 2.3347...$.

Good approximations to random fractals with D larger than two can be generated using computer graphics and a random number generator. These simulated mountain ranges are statistically self-similar and they do lead to very realistic looking pictures, even using a minimum of computer time. They are rapidly beginning to demonstrate the power of mathematics in computer art, and seem to suggest that impressive new artistic concepts can be developed through the mathematical modelling of the structure of pictures.

In fact, this computer modelling process can easily be formally extended to fractals of more than three dimensions. Although the eye cannot fully appreciate these structures, it is possible to extract three-dimensional projections (or slices) of these four-dimensional objects and thereby bring them into our field of view. This hardly does them justice—in the same way that someone who had never seen a cube could not fully appreciate its properties by looking at a sequence of two-dimensional pictures of it—but seems to be the best that we are able to accomplish in the way of pictorial representation.

The study of fractals is also providing useful sources of ideas in many scientific fields. Sponges, cloud formations, river catchments and particles of smoke may seem to have little in common, and yet each has recently been discussed in the scientific literature in terms of fractals. In the future they seem certain to have an impact on many diverse areas of science—particularly biology, geography and economics. Many research laboratories expend perceptible fractions of their entire basic-research budgets on the study of fractal systems, and whole research conferences are being devoted to the subject. And yet the whole enterprise is still in its infancy; the essential identity of fractals beyond their dimensional peculiarities has yet to be grasped. In many ways, as one eminent scientist has put it, the physics of fractals is a subject still waiting to be born. At present their major practical impact has been in computer graphics, particularly among groups who use fractals with ever greater frequency in the production of 'special effects' in the *Star Wars*-type of movie sequences.

15

The Road to Chaos

The world of mathematics, though incomprehensible to many, is generally believed to be a rather orderly domain. Precise questions are very carefully asked and precise answers are sought, and often obtained. It is true that there may be a troublesome problem of logic lurking here and there among the more abstract depths of mathematical thought, but the kind of arithmetic that we can carry out on our pocket calculators—that, in a sense, is easy. Every question has a 'well-ordered' answer. Even that peculiar number 0^0 of chapter 1, which confounded our calculator at first key-press, eventually succumbed to a little thought.

Life in the world of the pocket calculator should surely hold no terrors. Arithmetic is a very precise discipline and the calculator (or, to the more ambitious among us, the computer) is a friend. It is there merely to help us numerically when we are not able to solve some problem (maybe an equation or a set of coupled equations) in a more formal manner. Some equations are so easy, for those who have had even a little bit of background in elementary algebra, that assistance in solution from calculators is not necessary. For example, the equation $x^2 - 3x + 2 = 0$ would happily be solved by most of us using algebraic factorization to give the exact solutions $x = 1$ and $x = 2$. However, if we did not know any better method, we could still probably stumble upon the answers by merely 'testing' a few of the simpler possibilities.

On the other hand, with a slightly more difficult equation like $x^3 - x^2 - 3x + 3 = 0$, a calculator might prove useful (although those with a little better algebraic skill ought still to be able to find the exact solutions for x). The rest of us might first try testing a few easy whole numbers ($x = 1$ seems to work again) but the other answers, and there are two more, are not quite so obvious. This is where the calculator helps; it enables us to 'zero-in' on an

139

accurate numerical approximation even if we cannot solve the equation exactly. For example, trying a value $x = 1.7$ on the left-hand side leads to a right-hand side of $-0.0770...$, which is pretty close to the hoped-for value of zero. It therefore seems likely that $x = 1.7$ is close to another solution, but how close? Trying in turn the values $x = 1.71, 1.72, 1.73$ and 1.74 leads (via pocket calculator) to the sequence of right-hand-side numbers $-0.0538...$, $-0.0299...$, $-0.0051...$ and $0.0204...$. Since the latter numbers 'pass through' zero (that is, change from negative to positive) somewhere between $x = 1.73$ and $x = 1.74$, we can assume that there is another solution for x which is larger than 1.73 but smaller than 1.74. In addition, we can approach this solution ever more closely by introducing more and more decimal places; or at least until the capacity of our calculator is exceeded.

But how do we *know* that we can do this? Well, because simple mathematics is orderly. If you change the 'starting number' (in this case x) by a tiny amount, then you will only change the 'answer' on the right-hand side by a comparably tiny amount. In the present case this blind trust in the orderliness of arithmetic is fully justified, and we zero-in on a value $x = 1.732\,050\,80...$, which is the best my particular pocket calculator can do in approximating the exact answer $x = \sqrt{3}$. The third answer, by the way, is $x = -\sqrt{3}$.

Now, although we may suspect that the more dastardly of mathematicians could invent equations which, by cruel design, were less well-behaved near some special values of x (or even possibly a few which were badly behaved everywhere), we feel sure that these would not be 'simple' equations in the popular sense of the word, and this belief is not unreasonable. It was, in fact, very widely held among scientists and mathematicians until one day in 1960 when Edward Lorenz, of the Massachusetts Institute of Technology, programmed his powerful computer to numerically approximate solutions to a set of 13 coupled equations which modelled the Earth's atmosphere.

Although this problem was arithmetically complicated (and therefore required the numerical help of a computer), each of the individual equations involved was of an elementary kind; the kind which were believed to behave in an orderly fashion. Just as we did for the much simpler case above, Lorenz took a trial set of numbers which seemed to be quite close to the exact solution, and proceeded to try to zero-in on an even more accurate approximation. Changing his trial numbers by a tiny amount he set the computer to 'try again' and went off for a cup of coffee while he awaited the new results. When he returned he found that the computer's new output was not just a small refinement of the earlier one, but was completely different. His first suspicion was that the computer had malfunctioned, but on rechecking everything a horrible truth dawned; for this set of equations, tiny changes in the starting numbers produced wild fluctuations in the 'answer'. This was the shattering of a weatherman's dream. 'I knew right then', he said, 'that if the real atmosphere behaved like this, long range forecasting of weather was simply not possible'.

What Edward Lorenz had stumbled upon was the notion of chaotic behaviour. Chaos, in this sense, really means unpredictability. In the context of the meteorological equations it meant that a tiny change in the starting conditions (say a one degree change in temperature) could result in the difference between a day of sunshine or a day of torrential rain one month later. It implied that no degree of precision of meteorological knowledge (short of infinite precision concerning every relevant quantity) could achieve even an approximate degree of correct prediction sufficiently far into the future. In chaotic behaviour, even the slightest change in starting conditions becomes magnified to a degree which changes the final outcome beyond recognition. Moreover, the greater the difference between two sets of starting conditions the sooner this chaotic situation develops.

The distressing conclusion, since in practice all measurements have some finite degree of experimental precision, is that the final behaviour of any chaotic system can never be predicted, not even in a qualitative manner. Yet not quite all is lost. Lorenz found, when he asked his computer to print out thousands of possible solutions to his equations as he changed the starting values by extremely small amounts, that the output was not completely devoid of pattern. The numbers were random, it was true, but only within certain ranges of possible behaviour. Could there be, after all, some method in this madness? For example, although we must now accept that we cannot predict much about the weather one year from today, we can nevertheless set out many situations which seem to fall outside the 'allowed ranges' of possibilities. There do exist, after all, some qualitative aspects of weather which we all recognize; winters are cold, summers are warm, there are monsoon seasons and hurricane seasons and fairly well-defined limits of low and high temperature extremes. What is the source of this ghost of orderliness within chaos?

Before questions of this nature could be confronted it was necessary to find out exactly how simple mathematical equations strayed from the realm of orderliness to that of chaos, while still retaining the simplicity of form from which orderliness had come to be expected. The first task was to recognize what kinds of equations gave rise to chaotic solutions. The second was to study, if possible, the precise manner in which equations of this kind passed over from the realm of the predictable to the chaotic. One of the earliest clues came rather unexpectedly from a study in the mid-1970s of the yearly populations of seasonally breeding insects.

The equations which were thought to determine the changes of insect population from season to season were, unlike the analogous weather forecasting equations, simple in the extreme—pocket calculator simple! Suppose, for example, there was a stable population which, due to particular environmental criteria, numbered 500 000. The model then suggested that if you started off in the first season with a population of one million times x_1 (where $x_1 = 0.5$ would therefore be the stable population), then the resulting population in the second season would be one million times x_2 where

$$x_2 = 2x_1(1-x_1)$$

and the right-hand side means 2 times x_1 times $1-x_1$. Similarly, the population for the third season would follow according to

$$x_3 = 2x_2(1-x_2)$$

and so on down to season $n+1$, for which the population of one million times x_{n+1} would be related to the previous season's population of one million times x_n in exactly the same manner, namely

$$x_{n+1} = 2x_n(1-x_n).$$

The equation makes sense because, if in the first season there are too many insects for the environment to sustain (say 800 000, or $x_1 = 0.8$), then the second season will see some reduction to x_2 equal to 2 times 0.8 times 0.2, or $x_2 = 0.32$ (that is, 320 000 insects). With an underpopulation the conditions should now be favourable for population growth and, from the third season equation, x_3 equals 2 times 0.32 times 0.68, or $x_3 = 0.4352$ (that is, 435 200 insects), confirming our expectations.

Starting from season one with any x-value between 0 and 1 we can easily follow the predictions of the equations. Thus, from the above starting population having $x_1 = 0.8$, we quickly generate a sequence of values (that is, $x_1, x_2, x_3, x_4, ...$) as follows:

0.8, 0.32, 0.4352, 0.4916, 0.499 86, 0.499 999

which shows that by the sixth season we have reached the stable population. In fact, you will find that you can start with any x_1 you like between 0 and 1 and always finish up with a stable limit of $x = 0.5$. What could be simpler than that! The stable value (in our case $x = 0.5$) is sometimes called a 'fixed point' and, since all roads seem to lead to it, it is also called a 'stable fixed point', or an 'attractor'. This is important because the above set of equations has another fixed point which may have escaped your notice. It is $x = 0$. If you start with $x_1 = 0$, then the sequence of numbers generated for $x_2, x_3, x_4,$ etc are all zero. This is the mathematical way of saying that, if you start with no insects at all, you will be extremely fortunate to breed any, in spite of how many seasons you wait. The value $x = 0$ is therefore also a fixed point. However, it is an unstable fixed point in the sense that no matter how small a non-zero value you choose for x_1, the sequence of values x_n will now always move away from $x = 0$ towards its attractor fixed point $x = 0.5$; try it and see.

So the equations are at least not nonsensical; but they are also very far from chaotic. In fact they are extremely 'well-behaved', zeroing-in on 0.5 in a most orderly of fashions. But now let us suppose that the stable population is one million times 0.6875 (that is 687 500 insects). There is nothing magic about this number; I have picked it solely because it is larger than $x = 0.5$ (which is the particular direction I wish to investigate) and it has a general generating equation for seasonal fluctations which is easy to write down. It is

$$x_{n+1} = 3.2x_n(1-x_n)$$

and is changed from the earlier one only in replacing the multiplying factor 2 on the right-hand side by 3.2. If you take your pocket calculator and start with $x_1 = 0.6875$, you will quickly find that the entire generated sequence x_2, $x_3, \ldots, x_n \ldots$ has the same value 0.6875. This then is indeed a fixed-point solution. If you start with 687 500 insects then you will have that number forever. But is $x = 0.6875$ an attractor? Well, check it out for yourself. If you do you will find that the sequence of x_n-values begins to oscillate back and forth between two values, one near 0.799 and the other near 0.513. On my calculator, which can manage nine decimal places, the sequence finally settles down to a form

$$a, b, a, b, a, b, a, b, \ldots$$

where $a = 0.799\,455\,490$ and $b = 0.513\,044\,510$. What on earth has happened? Quite evidently, the fixed point $x = 0.6875$ is an unstable one. The stable situation describes a repetitive fluctuation between an overpopulated season with 799 455 insects (neglecting fractions of an insect to retain credibility) and an underpopulated one with 513 044 insects. Physically, this seems quite possible. Our equations have not led to an absurdity, but merely to a situation which is a little more complicated than we perhaps expected.

However, worse is yet to come! As we increase the value of our fixed point further, our so-simple equations begin to do the most incredible things. With a fixed point of $x = \frac{5}{7}$ our governing equation becomes

$$x_{n+1} = 3.5x_n(1-x_n)$$

and you can quickly check out that a population of $x = \frac{5}{7}$ would happily reproduce itself forever. But what if we start off from another value of x (in fact *any* other except the trivial fixed point $x = 0$ which is present in all these systems)? Try it again on your calculator. Once more we find that the non-trivial fixed point (this time $x = \frac{5}{7}$) is not an attractor, but now the stable solution involves four numbers, rather than two, and the sequence settles down to a pattern

$$a, b, c, d, a, b, c, d, a, b, c, d, \ldots$$

with (again to my nine-decimal-place accuracy) $a = 0.382\,819\,683$, $b = 0.826\,940\,707$, $c = 0.500\,884\,210$ and $d = 0.874\,997\,264$. Once more we can envisage the physical situation. In season 'a' there is an underpopulation which leads to an increase in numbers in the following season 'b'. But this has now produced an overpopulation which therefore leads to a decrease in season 'c'. Finally, 'c' is once more underpopulated and triggers the final increase to season 'd'. Without actually doing the calculation, you would probably expect that such an oscillating behaviour would eventually approach more and more closely the fixed-point value of $x = \frac{5}{7}$; but it never does—the oscillating values themselves are perfectly stable.

Thus, as the numerical constant in the generating equation has gone from 2 to 3.2 to 3.5, the character of the stable limit has changed dramatically from a stable fixed point, to an oscillating 'two-cycle', to a repeating 'four-cycle'. One is tempted to ask how long this pattern of increasing complexity can go on? The answer, in a sense, is forever. You see, as the numerical constant increases beyond the value 3.5 last considered, the number of points in the stable cycling pattern begins to explode, through 8, 16, 32, 64, ... to ever larger powers of 2 as the 'equation constant' nears a special (so-called 'critical') value close to 3.57. Beyond this critical value the situation becomes completely chaotic. By this we mean that the final pattern of numbers never repeats itself; in a sense it is cycling with an infinite number of points in the cycle pattern.

Thus, our simple defining equation

$$x_{n+1} = Ax_n(1 - x_n)$$

has led us from a most orderly domain of mathematics, through increasingly complex (though still stable) cycling configurations, to completely chaotic behaviour—and all by a simple steady increase in the parameter A from 2 to about 3.6. For values of A larger than 3.6 the chaotic behaviour is maintained, but with the degree of chaos steadily increasing. To see what we mean by 'degree of chaos' it will be necessary to examine at least one chaotic situation (i.e., using a value of A larger than the critical value in the defining equation) in some detail.

One of the most fearful properties of chaos is that any two starting concentrations x_1, even those chosen arbitrarily close in value to each other, eventually move apart to patterns which have no inter-relationship whatsoever. Let us examine this particular chaotic property using our insect-breeding defining equation, with $A = 3.9$. In particular, we shall consider the fate of two insect populations with the closely equal starting values of $x_1 = 0.100$ and $x_1 = 0.101$ (representing 100 000 and 101 000 insects) respectively. The following columns of numbers (given only to a three-place decimal accuracy but actually calculated to a nine-place precision by my trusty pocket calculator) compare the fortunes of these two respective starting populations when governed by that innocent looking equation above through the first 14 generations:

Generation	Set A	Set B
1	0.100	0.101
2	0.351	0.354
3	0.888	0.892
4	0.387	0.376
5	0.925	0.915
6	0.271	0.304
7	0.771	0.825
8	0.690	0.562

(cont)

Generation	Set A	Set B
9	0.835	0.960
10	0.538	0.150
11	0.969	0.497
12	0.116	0.975
13	0.399	0.095
14	0.935	0.336

It is seen that by the 10th season, any semblance of correspondence between the two sets of populations has completely disappeared.

There is nothing at all magical about the particular defining equation which we have used to set up this mysterious progression from orderliness to chaos. Countless other simple relationships can be written down which produce this same kind of behaviour. In fact, it is this uniformity which makes a study of the onset of chaos so fascinating. The common feature is the passage from a stable fixed point, through an endless sequence of cycle doublings, to chaos, as some equation parameter A gradually approaches its critical value (say at $A = L$, for Limit). The farther beyond the limiting value L the parameter A progresses, the fewer the number of generations required to lose all correlation between two sets of nearly equal starting 'populations'. This enables us to differentiate between different degrees or strengths of chaotic behaviour in an obvious fashion.

The most impressive display of universal behaviour is seen on approach to the critical limit $A = L$ from the 'ordered' side, which exhibits the repeated cycle-doublings. Each of these doublings (or 'bifurcations', as the mathematicians in their endless pursuit of simplicity prefer to call them) occurs when the parameter A reaches special values; say $A = A_n$ for the doubling to a cycle with 2^n repeating values. Eventually this sequence of special A-numbers $A = A_n$ converges (as n progresses to infinity) on the limiting value $A = L$ for which chaos sets in. It follows that the number $L - A_n$ gets smaller and smaller as n gets larger and larger. Quantitatively, it is found that the ratio

$$\frac{A_{n+1} - A_n}{A_{n+2} - A_{n+1}} = d_n$$

very quickly approaches a strange but precisely defined number as n increases. This number is

$$d = 4.669\ 201\ 6\ldots .$$

What is truly amazing is that this number d has *exactly* the same value for *every* model which exhibits a cycle-doubling route to chaos. This also implies that $L - A_n$ becomes proportional to d^{-n} as n becomes large, with d again equal to the value set out above. Chaos therefore sets-in in a remarkably unchaotic manner—one, in fact, which is very precisely defined. At the onset of chaos all details of the specific defining equation seem to become irrelevant and a kind of universal behaviour takes over.

But what does all of this have to do with real chaotic motion in a scientific sense? Well, it turns out that a rather wide range of real physical, chemical, and biological systems actually do follow a cycle-doubling route to chaos. These systems include the onset of turbulence in gases (a common example of which is the behaviour of smoke rising from the end of a cigarette), electrical instabilities in electronic circuits, and even the response of heart muscle to a pacemaker. What is even more impressive is the fact that experimental efforts to measure the d-value as defined by the ratio given above do appear, in the different scientific contexts, to point to a universal constant with a value close to that of 4.669 ... which 'falls out of' the mathematical theory.

Impressive though these findings now appear, the new and seemingly rather bizarre ideas involved in the cycle-doubling road to chaos did not, at first, find willing acceptance among the mathematical Moguls of the academic world. The original demonstration (essentially just that set out in this chapter) was first given for the 'insect population' equations by a young physicist at Los Alamos using a pocket calculator. His name was Mitchell Feigenbaum, and his papers discussing the derivation of the universal constant d (which is now called Feigenbaum's constant, in his honour) were for three years, from 1976–1979, consistently rejected by the editors of professional mathematical journals. After all, what of serious mathematical consequence could possibly result from doodling on a pocket calculator in this enlightened age? Only in 1979, when an Italian physicist (who was attempting to solve by computer five messy equations which modelled real-life turbulence) observed the same cycle-doubling phenomenon, was Feigenbaum's work finally taken seriously.

It then led to physical experiments which looked for, and found, the cycle-doubling process in nature. The first such successful experiment took place in France, and consisted of heating a bath of liquid helium until it boiled into chaos. As the experimenters carefully observed the helium liquid approaching 'the boil', they spotted convection currents begin to swirl, and then to go through what were undoubtedly successive episodes of doubling. This pioneering work was quickly followed by other confirmations. There was indeed order along the road to chaos! One of the more recent and most fascinating applications has been an effort to model the heart's passage from rhythmic life sustaining beats to fatal spasms—a heart attack. Equations were derived which were able to reproduce a wide range of observed heart-rhythm disturbances. And again, before the disturbances set in, the equations exhibited unmistakable signs of a period-doubling route to heart-beat chaos.

Although it is now known that cycle-doubling is not the only route to chaos, it does remain the classic example. As a result of Feigenbaum's pioneering work, serious study of the onset of chaos is now an active scientific discipline. Scientists in a variety of fields are rushing to make up for lost time. After all, chaos is everywhere, from the microworld of quantum physics to the macroworld of astrophysics. A scientific revolution is upon us, and to think that the spark was provided by the humble pocket calculator. How rapidly the new subject will develop, and to what extent its full promise will be fulfilled, remains to be seen. However, it is important to bear in mind that what is being studied is the degree of orderliness of the onset of chaos. The chaotic state itself is still what the word implies. It means that, since an infinite degree of accuracy is experimentally unattainable, the results of a 'well-controlled' experiment need not always be reproducible. From the weatherman's standpoint, although a glance at the sky may still be sufficient to assess the weather for the next hour or so, and a careful monitoring of global weather statistics may enable tomorrow's forecast to be presented with some degree of confidence, long range forecasting is likely to forever remain a thankless task.

16

Super-mathematics and the Monster

Examples of symmetry are all around us. Why, we ourselves are approximately symmetric about a vertical line (or more correctly a vertical plane) down our middle. In fact, so many objects around us have approximately this same so-called bilateral symmetry that, when we look in a mirror, it is not at all obvious that in the mirror world right has been changed into left, and vice versa. Why mirrors should only invert left and right (and not up and down) is itself a question which, should you choose to think deeply about it, will probably cause you a few sleepless nights. But this is not the question which is to be pursued in this chapter, so that I (rather unkindly) will say no more about it here. Some objects, of course, have other kinds of symmetry associated with them. The swastika, for example, possesses some sort of rotational symmetry, while a star has both a rotational symmetry and a bilateral one. Wallpaper, on the other hand, has a somewhat different kind of symmetry where the whole pattern can be displaced in various directions without looking any different.

Let us now ask ourselves a little more precisely what we mean by symmetry, and then try to answer the question as to what objects with the same symmetry really do have in common. Symmetry is defined in terms of an 'operation' (which might be a reflection, a rotation, or a displacement) after the performance of which the object in question 'still looks the same'. As a simple example consider a square. Obviously, it can be rotated through 90 degrees (clockwise or anti-clockwise) or through 180 degrees and 'still look the same'. These rotations are then referred to as symmetry operations of the square, but they are not the only ones. Clearly, we can reflect the square through either of its two diagonals, or either of two lines passing through its centre parallel to two of its sides. This makes seven operations and finally, if we add the 'stay as you are' operation, that makes eight.

We say that there are eight symmetry operations under which a square remains invariant (which is the mathematical word for 'the same'). But, you may ask, what if I rotate the square through 270 degrees or 360 degrees, or turn it over? Are these not additional operations? The answer is no, and to make this clear we draw a point inside the square at any position which is not special (say just a little way down one edge from a corner). All operations which move this point to a different position but leave the rest of the square 'invariant' are different symmetry operations. Any two seemingly different operations which move both the square and the point in the same way are, from a symmetry point of view, the same.

With this out of the way we can say that the property which all squares have in common is the identity of its symmetry operations, but it goes deeper than this. You see, any two operations performed one after the other are always equivalent to another of the symmetry operations. In this sense the operations form a 'closed' set and if we label them by, say, the first eight letters of the alphabet, we can take a checkers board and, labelling the rows A,B,C,D,E,F,G,H and the columns in the same way, fill in the squares in a manner which will tell us exactly what happens with these square operations when we combine them. Thus, in row E and column F we write the operation which results from first doing operation E then operation F. Since the set is closed, the whole checker board will be filled with a special pattern of letters between A and H. This is called the 'multiplication table' for the symmetry operations of a square. The pattern tells us something more about the square, but what use is it?

To answer this question let us consider the very simplest symmetry of all, the (approximate) symmetry of ourselves. This symmetry group (for the term 'group' is given to sets of operations of this kind) has only two operations; the 'stay as you are' one, which we shall write as I (for identity), and the reflection 'through the plane down the middle', which we write as R (for reflection). If we use our wristwatch as the arbitrary point, the reflection takes it from our left wrist to our right one, or vice versa if we are left-handed. Any other operation, like turning a somersault, does not count as a new operation; it is still 'operation I' since it leaves the wristwatch exactly where it was before the event (assuming it did not come off!). Our own symmetry group is therefore composed of the two operations or 'elements' I and R.

We can now write down its multiplication table. Since 'stay as you are' followed by reflection is the same as reflection, we write in symbols that $IR = R$. Similarly, we must also have $RI = R$, which is just the symbolic way of saying that reflection followed by 'stay as you are' is also the same as reflection. Finally, since one reflection followed by a second takes the wristwatch back to where it started, we have $RR = I$. These four equations can be represented on a 2 by 2 multiplication-table checkers board in the manner

	I	R
I	I	R
R	R	I

We now notice that it is possible to replace I and R by numbers and still obey the multiplication table if we interpret IR arithmetically as 'I times R', and so on. Moreover this can be accomplished in two separate ways; firstly, by putting $I = 1$, $R = 1$, and secondly, by putting $I = 1$, $R = -1$. Each of these choices satisfies the four equations $II = I$, $IR = R$, $RI = R$ and $RR = I$ as required. We say that we have found two 'representations'; $I = 1$, $R = 1$ being called the symmetric representation and $I = 1$, $R = -1$ the anti-symmetric one.

Where has this got us? Well, it tells us, for example, that any motion which takes place subject to forces which possess this I,R kind of symmetry can be of only two fundamentally different types; symmetric or anti-symmetric. In order to understand this clearly it is only necessary to choose a simple example. In figure 20(a) we show three beads on a piece of straight wire. The outside beads are of the same mass (say m) while the centre bead is of a possibly different mass (M), and each outside bead is connected to the centre one by identical springs. It is clear from the figure that, for this system, the plane through the centre bead perpendicular to the wire is a plane of reflection symmetry. This arrangement is therefore subject to the symmetry restrictions spelled out above.

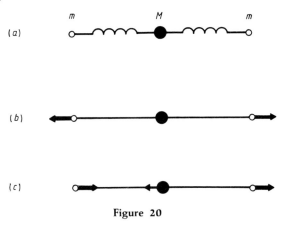

Figure 20

Suppose now that we gently disturb the beads and set them into oscillatory motion. The above 'group theory' tells us that there are only two possible sorts of 'simple' motion, by which term I mean motion that repeats itself endlessly in simple oscillations (ignoring frictional slowing down). These so-called 'normal modes' are the symmetric and anti-symmetric modes illustrated in figures 20(b) and 20(c) respectively. But how do we know that these are the relevant modes, and in what sense are they 'symmetric' or 'anti-symmetric'?

Let us look first at figure 20(b). In it, the arrows indicate the displacements of the masses at a particular moment of time. The end masses m are moving equally outward, and the centre mass M is stationary. Later on in the oscillation, the end masses will move equally inward with M still at rest. This motion is the symmetric mode because it has the same properties under I and R transformation as the symmetric representation $I = 1$, $R = 1$. For example, when figure 20(b) is reflected through the reflection plane it turns into itself; i.e., the operation R is equal to 1. The operation I, of course, meaning 'stay as you are', is always equal to 1. Note that in this symmetric mode the outside masses m must always be moving in an *exactly* 'opposite' fashion, while the centre mass M must be *exactly* at rest. Any motion of M would violate the symmetric-mode symmetry requirement of $R = 1$.

Now let us look at figure 20(c); this is the anti-symmetric mode. In this mode the outside masses move equally in the same direction, and the centre mass can now also move. In fact, if we wish to keep the 'centre of gravity' of the entire mode fixed, then the centre mass must move in a direction opposite to that of the outside masses. And what happens if we reflect this mode in the reflection plane? It is easily seen from the figure that it reflects into its negative (that is, the arrows indicating displacement all turn around to the opposite direction). This means that, for this particular mode, the operation R is equivalent to -1. Again, with $I = 1$, we now identify this mode as conforming to one of the allowed representations—this time the $I = 1$, $R = -1$ anti-symmetric one.

Now, in truth, if you set up the arrangement shown in figure 20(a) and you begin by just tweaking one of the end beads, the resulting motion will appear to be quite different from (and much more complicated than) either motion shown in figures 20(b) and 20(c). However, this motion (and any other which you can contrive by starting off in other ways) can always be described as a mixture of these two 'normal modes' in different proportions. In this sense, the symmetric and anti-symmetric normal modes are the fundamental modes from which all possible motions can be made up and their basic structures we know solely from the symmetry of the set-up. It is true that symmetry alone does not tell us all we might wish to know (the individual frequencies of the two normal modes for example; these depend on the detailed physics of the problem) but, amazingly, the fundamental *pattern* of each normal mode motion is completely determined by 'group theory', that is, by the manner in which the symmetry operations combine.

Exactly the same sort of reasoning can be carried out for other symmetry arrangements, such as triangles, squares, rectangles and countless other shapes. For each, we can locate all the symmetry operations and work out the multiplication table and arithmetic representations. Although these representations do get more complicated as the number of symmetry operations (called the 'order' of the group) increases, and often require the use of more sophisticated mathematical concepts like complex numbers, they exist and often tell us some amazing things. Sometimes it is that two seemingly quite different problems are, from a symmetry point of view, identical, i.e., have multiplication tables which are identical even though the physical meaning of the symmetry operations is different. At other times it is that two or more representations (or normal modes) are specially related (turning into each other under symmetry operation) and must therefore all have the same 'frequency' regardless of the physics of the situation.

The same ideas can be applied to solid figures, like the cube, for example, which has 48 symmetry operations (24 involving rotations and 24 involving reflections). They can also be extended to include symmetry problems which concern lateral movement as well. For example, there are only 17 different symmetries of wallpaper. The equivalent problem in three dimensions is more formidable, there being no less than 230 possible symmetries which atom arrangements in crystals (for example) can have. They are called 'space groups' and play an extremely important role in determining and classifying the atomic structure of solids like table salt (sodium chloride) or iron metal.

Experimentally, scientists determine atomic structures by focusing beams of X-rays or neutrons onto a sample and observing the way in which the beam is split up in different directions (scientists say 'is diffracted') on passage through the atomically ordered crystalline lattice. The special directions in which the diffracted beams emerge (and equally importantly those symmetry directions in which diffracted beams are rigorously absent) are determined by the space group of the crystal. Careful measurement of the directions and relative strengths of the diffracted beams therefore finally enables the complete atomic structure to be obtained. The known restrictions imposed by 'space-group' symmetry simplify the task of identifying the atomic positions enormously.

All the geometric symmetries so far discussed have concerned operations involving only discontinuous 'jumps' of some kind; like a rotation of 60 degrees or a lateral movement of some fixed distance. There are other, so-called 'continuous', symmetry groups which can be related to the symmetry of the circle (in two dimensions) or the sphere (in three). These figures remain unchanged even after rotation operations of infinitesimal size. They are of obvious interest to the scientist in connection with systems which possess axial symmetry or spherical symmetry respectively.

Consider, for example, the case of the hydrogen atom. It is composed of a heavy nucleus (or point-sized mass as far as we are concerned) orbited by a single electron. The electron moves subject to motional equations which must contain within them the symmetry restrictions of spherical space. Although the mathematical details for this situation are beyond the scope of a book like this one, the possible orbits are just as clearly defined as were the symmetric and anti-symmetric modes of the three-bead problem discussed earlier. Unfortunately for most of us (who are not expert scientists) the electron cannot truly be pictured as a tiny particle, but more as a 'cloud' surrounding the nucleus. The symmetry restrictions therefore show up in the allowed shapes for this cloud. The simplest shape is one with full spherical symmetry, and scientists call it an s state (although, oddly enough, the s here stands neither for spherical nor for symmetric; but that is another story). Others include clouds of dumb-bell-shaped lobes in increasingly complex orientational configurations, but all precisely defined by the continuous symmetry group of the sphere.

Yet the enormous power of 'group theory' goes far beyond anything we have so far mentioned. To this point we have associated the group operations with the rearrangement of actual objects, like triangles, squares, cubes, spheres, and crystal structures. This is the realm of the scientist. Real mathematicians allow their minds to reach beyond the physical world. Many other symmetries, in addition to geometric ones, are 'out there' waiting to be discovered. Consider, for example, the clock arithmetic discussed in chapter 7. In performing addition sums on an ordinary 12-hour clock we note results like $8+2 = 10$, $8+6 = 2$, $6+9 = 3$, $6+6 = 0$, etc. Under the operation of 'addition' the numbers form a closed set (no new numbers appear on the right-hand side) and zero performs the function of 'stay as you are'. For a three-clock (instead of a 12-clock) we can easily write down all of the possible addition sums as follows:

$$0+0 = 0 \qquad 0+1 = 1 \qquad 0+2 = 2$$
$$1+1 = 2 \qquad 1+2 = 0 \qquad 2+2 = 1.$$

If we now let the numerals 0, 1, 2, play the role previously reserved for symmetry elements, and let addition be adopted as the manner of combining them, then we quickly obtain the 'multiplication' table

	0	1	2
0	0	1	2
1	1	2	0
2	2	0	1

We can rewrite this in letter form by noting that 0 plays the part of the identity element ($0 = I$), and arbitrarily redefining $1 = A$ and $2 = B$. Correspondingly, we obtain

	I	A	B
I	I	A	B
A	A	B	I
B	B	I	A

You should now find it very easy to verify that this is exactly the same multiplication table that results from examining the symmetry operations of the emblem of the Isle of Man, which is three legs, 120 degrees apart, kicking each other around the central point where they join (in other words a figure with a threefold rotational symmetry but no reflection symmetry).

Now we have something interesting. The *same* group multiplication table has arisen in two quite different contexts. Could it be that all order-three groups of operations have the very same table and, therefore, that this 'group' has an absolute significance over and above any particular example of 'order-three symmetry' which we may have used to model it? The answer is yes, and the same is true of the order-two group discussed earlier. Are all order-four groups also the same? If you check out the 'multiplication' tables for a four-clock and a rectangle you will find that this time the answer is no, but there are only two different groups of order-four and the two examples cited generate them.

Upon analysing multiplication tables of higher order it becomes apparent that these 'objects', though obviously closely related to 'symmetry' in some general sense, do indeed have a reality over and above any specific context. When looking at these tables the mathematician sees only the pattern. To him the structure of the group of operations is the only important thing. The nature of the operations or elements is not important, but only the manner in which they interlock. With this in mind it is possible to define a 'group' in purely abstract terms and to create a kind of super-mathematics which is reflected in the patterns of the multiplication tables. There are only four rules which define this super-mathematics called 'group theory'.

(1) A 'group' is a set of 'elements' $A, B, C, \dots$ etc, which can be combined with one another to get other elements. The combination of any two elements AB is also a member of the group.

(2) One of the elements I is a 'stay as you are' or identity element, so that $IA = AI = A$ for all elements A in the group.

(3) For each element A in the group there is another A^{-1} such that $AA^{-1} = A^{-1}A = I$.

(4) For all elements A, B, C in the group $A(BC) = (AB)C$, where the brackets mean 'do that combination first'.

It is now very easy for you to check out that these purely abstract rules are indeed obeyed by the specific examples we looked at earlier. Using these rules without any association of the elements A, B, C etc with actual physical operations, the 'super-mathematician' is now able to deduce how many groups (that is, different multiplication table patterns) there are for any particular order. The list begins as follows:

Order	Number of groups
1	1
2	1
3	1
4	2
5	1

(cont)

Order	Number of groups
6	2
7	1
8	5
9	2
10	2
11	1
12	5

Each group is a unique mathematical object with its own 'personality', and people who work in group theory get quite chummy with them and give them special names. Their ultimate significance, however, goes even deeper than this because, just as in arithmetic every integer possesses a unique factorization into a product of primes (for example, $60 = 2 \times 2 \times 3 \times 5$), so every group can be 'factored' in a certain sense. Some groups have smaller groups inside them in such a way that they can be expressed as the 'product' of these smaller groups. The 'prime numbers' of group theory are therefore those groups which can only be factored into themselves and the single identity element I. These are the building blocks of group theory. They are called the 'simple groups of finite order' and an effort to classify them completely has been underway for well over 100 years and has, in the mid-1980s, just been completed.

The sheer immensity of the task can be judged from the fact that the final proof occupies upward of 10 000 printed pages scattered over some 500 articles in technical journals. Most of the work was done between the the late 1940s and the early 1980s and has involved the efforts of more than 100 mathematicians. Even when it is finally condensed and refined it is still expected to occupy well over 1000 pages of text.

There are, perhaps not surprisingly, an infinite number of simple groups of finite order, in the same way that there are an infinite number of prime numbers. However, the group theoretical problem is more complicated because, not only are there several different kinds of infinite family, there are also certain renegades (called 'sporadic' groups) which do not fit into any of these infinite families. The simplest example of an infinite family is that of the clock-number groups with a prime number of elements; in other words the n-clocks with $n = 2, 3, 5, 7, 11, 13$, etc. The other examples of infinite families of simple groups are much more difficult to classify in detail, but eventually 17 more such families were constructed and it is now known that these 18 families complete the list.

The first five of the puzzling sporadic simple groups had been found as long ago as the 1860s. The smallest of these was of order 7 920 and the other four of orders 95 040, 443 520, 10 200 960, and 244 823 040 respectively. Quite evidently they were far from trivial entities and these five remained the only known sporadic simple groups for a whole century.

Then, in 1966, Zvonimir Janko, then at Monash University in Australia, discovered a sixth, thereby launching what might be called the modern era of sporadics. Thereafter these strange new sporadic creatures began to pop up at the rate of about one a year into and through the 1970s. The size of them was impressive indeed, many having more than 10^{10} elements. The climax came in 1982 when Robert Griess Jr, of the Institute for Advanced Study in Princeton, constructed a sporadic group which was enormously larger than any other yet found. It came to be known as 'the monster' because of its size. It possessed no less than 808 017 424 794 512 875 886 459 904 961 710 757 005 754 368 000 000 000 elements, which is about 8×10^{53}, and therefore had more than 6×10^{107} places in its multiplication table. It is now known to be the largest sporadic group which exists and must, of course, represent the symmetry properties of something pretty impressive. But what? It is, in fact, intimately connected with the symmetry properties of that extremely dense 24-dimensional laminated lattice packing L_{24} discussed in chapter 13.

Ultimately, 26 sporadic groups were discovered, with the monster being the largest. The second largest, the 'baby monster' has about 4×10^{33} elements which, although it still sounds pretty large (and is), is a factor of more than 10^{20} smaller in order than the monster itself. The smallest sporadic is the one mentioned earlier with 7 920 elements. The final proof concerning the complete classification of *all* the finite simple groups now claims that only 26 sporadic examples exist; just 26 maverick groups which evade classification as part of infinite families. These, together with the 18 infinite families referred to earlier, therefore form the ultimate building blocks of group theory or, in other words, of symmetry in its most general and abstract form.

So, in a sense, the theory of groups has reached maturity. But where did it all start? The invention of symmetry groups is generally credited to the youthful genius Evariste Galois, a French mathematical prodigy, who died in his 21st year by a pistol shot received in a senseless 'duel of honour'. Born in 1811, he was younger than Gauss by 34 years, although the venerable Gauss outlived him by almost a quarter of a century. Galois was barely 17 when he began to create that branch of mathematics which now provides insights into symmetries all the way from particle physics to Rubik's cube. His most famous work centred on an effort to understand why great mathematicians who had preceded him had managed to find general solutions to only a very few specific kinds of algebraic equation.

Algebraic equations are ones like

$$x + 2 = 5$$
$$x^2 - 4x + 11 = 0$$
$$x^3 + 2x^2 - 6x - 2 = 0$$
$$2x^4 + x^3 + 5x - 7 = 0$$
$$3x^5 - x^4 - 5x^3 + 2x^2 - 7 = 0$$

which contain various integer powers of the 'unknown' x, all with integer coefficients (some of which may, of course, be zero). The largest power involved is called the 'degree' of the equation, so that the equations above are examples of degrees one through five in order. The most general example of an algebraic equation of a particular degree (say four) can be written

$$Ax^4 + Bx^3 + Cx^2 + Dx + E = 0$$

where A, B, C, D, and E can stand for *any* integers. The question to be answered is 'can this equation be solved in the form $x =$ (some expression involving A, B, C, D, E connected by the basic operations of arithmetic like $+$, $-$, $\div$, $\times$, powers and roots)?' And, in Galois' day, the answer for the most general equations of degree one through four was known to be yes. However, for degrees five and larger the answer was not known.

The general answers for degrees one and two had been known since Babylonian times and are contained in all high school algebra books. The general solutions for 'cubic' (degree-three) and 'quartic' (degree-four) equations had been obtained in the 16th century. When Galois approached the problem in the 1820s, nearly 300 years of fruitless effort had been made by generations of great mathematicians to 'solve' the general 'quintic' (degree-five) equation.

Galois wrote his first paper on calculating solutions of fifth-degree equations when he was 17 years old, and over the following two years he derived the concept of 'groups'. This enabled him to set out in a systematic fashion the relationships which the solutions of various kinds of algebraic equations must have with one another by virtue of the symmetry properties of the equations themselves. The groups which he derived (now called Galois groups in his honour) showed clearly that no equation of degree larger than four possessed a general solution of the form sought. Three hundred years of effort to solve the general quintic had failed for the very good reason that no solution existed. Solutions do exist, of course, for *particular* quintic equations (with A, B, C, D, E set equal to specific numbers) but these solutions have to be obtained numerically rather than algebraically.

Using the concepts of symmetry groups Galois was not only able to 'dispose of' the problem for quintic equations; the power of the method was such that it could attack the equivalent problem for general algebraic equations of *any* degree. The power of 'symmetry' was dramatically demonstrated for the first time. And although Galois-group symmetries are not directly related to geometry, they are just as precisely determined as are rotations and reflections. However, he was so far ahead of his time that the full significance of his work took many more decades to be properly recognized. One can only imagine by how much this period would have been shortened had Galois himself survived to live out a more normal span of years. Algebra, in a sense, was set free by Galois. The focus moved from specific problems to generalizations and abstractions. It ushered in the era of

'modern algebra' or 'abstract algebra'; a super-mathematics in which operations are as abstract as the quantities they operate on. Only the shapes
remain—much like the smile on the Cheshire cat—shapes which make up the
theory of groups and which lead inevitably to The Monster.

Bibliography

A selection of books for further reading.

BARNETTE D W 1984 *Map Coloring, Polyhedra and the Four Color Problem* (Washington, DC: Mathematical Association of America)

BARNSLEY M F 1988 *Fractals Everywhere* (San Diego, CA: Academic)

BROOK R J 1986 *The Fascination of Statistics* (New York: Marcel Dekker)

CONWAY J H and SLOANE N J A 1988 *Sphere Packings, Lattices and Groups* (Berlin: Springer)

CROWE D K and CROWE D W 1988 *Symmetries of Culture* (Seattle, WA: Washington University Press)

EYNDEN C V 1986 *Elementary Number Theory* (New York: Random House)

GLEICK J 1987 *Chaos, Making a New Science* (New York: Viking)

GRUNBAUM B and SHEPHARD G C 1986 *Tilings and Patterns* (San Francisco: W H Freeman)

HOGGATT V E 1969 *Fibonacci and Lucas Numbers* (Boston, Mass.: Houghton-Mifflin)

LOWRY R 1989 *The Architecture of Chance* (Oxford: Oxford University Press)

MANDELBROT B B 1982 *Fractal Geometry of Nature* (San Francisco: W H Freeman)

ROSEN K H 1988 *Elementary Number Theory and its Applications* 2nd edn (Reading, Mass.: Addison Wesley)

ROSENFELD B A 1988 *A History of Non-Euclidean Geometry* (Berlin: Springer)

RUCKER R 1977 *Geometry, Relativity and the Fourth Dimension* (New York: Dover)

——1984 *The Fourth Dimension: Toward a Geometry of Higher Reality* (Boston, Mass.: Houghton-Mifflin)

VAN TILBORG H C A 1988 *Introduction to Cryptology* (Norwell, Mass.: Kluwer)

WELSH D 1988 *Codes and Cryptography* (Oxford: Oxford University Press)

Index